NATURE AND THE GREEKS

and

SCIENCE AND HUMANISM

NATURE AND THE GREEKS

and

SCIENCE AND HUMANISM

Erwin Schrödinger

With a foreword by
Roger Penrose

CAMBRIDGE
UNIVERSITY PRESS

University Printing House, Cambridge CB2 8BS, United Kingdom

Cambridge University Press is part of the University of Cambridge.

It furthers the University's mission by disseminating knowledge in the pursuit of education, learning and research at the highest international levels of excellence.

www.cambridge.org
Information on this title: www.cambridge.org/9781107431836

Nature and the Greeks
First published 1954

Science and Humanism
First published 1951
Reprinted 1952, 1961

First published 1996
Canto Classics edition 2014
4th printing 2021

Printed in Great Britain by Ashford Colour Press Ltd.

A catalogue record for this publication is available from the British Library

ISBN978-1-107-43183-6 Paperback

CONTENTS

Foreword by Roger Penrose vii

o

NATURE AND THE GREEKS

I The motives for returning to ancient thought 3
II The competition, reason *v.* senses 22
III The Pythagoreans 34
IV The Ionian Enlightenment 53
V The religion of Xenophanes. Heraclitus of Ephesus 69
VI The Atomists 75
VII What are the special features? 90
Bibliography 99

o o

SCIENCE AND HUMANISM

Preface 103
The spiritual bearing of science on life 105
The practical achievements of science tending to obliterate its true import 113
A radical change in our ideas of matter 115
Form, not substance, the fundamental concept 122
The nature of our 'models' 125
Continuous description and causality 130

The intricacy of the continuum 133
The makeshift of wave mechanics 143
The alleged break-down of the barrier between subject and object 151
Atoms or quanta – the counter-spell of old standing, to escape the intricacy of the continuum 157
Would physical indeterminacy give free will a chance? 162
The bar to prediction, according to Niels Bohr 168
Literature 172

FOREWORD

oo

by Roger Penrose

I vividly recall reading Erwin Schrödinger's slim volume *Science and Humanism* some forty years ago, probably at a time while I was still a research student in Cambridge. It had a powerful influence on my subsequent thinking. *Nature and the Greeks*, although based on slightly earlier lectures, was not published until somewhat later, and I have to confess that I did not come across it then. Having only now read it for the first time, I find a remarkable work, of a similar force and elegance.

The two volumes go well together. Their themes relate closely to each other, being concerned with the nature of reality and with the ways in which reality has been humanly perceived since antiquity. Both books are beautifully written, and they have a particular value in enabling us to share in some of the insights of one of the most profound thinkers of this century. Not only was Schrödinger a great physicist, having given us the equation that bears his name – an equation which, according to the principles of quantum mechanics, governs the behaviour of the very basic constituents of all matter – but he thought deeply on questions of philosophy, human history and on many other issues of social importance.

In each of these works Schrödinger starts by discussing pertinent social issues concerning the role of science and of scientists in society. He makes it clear that, whereas there is no doubt that science has had a profound influence on the modern world, this influence is by no means the real reason for doing science; nor is it clear that this influence is itself always positive. However, his main purpose is not just to discuss issues of this kind. He is primarily concerned with the very nature of physical reality, of humanity's place in relation to this 'reality' and with the historical question of how great thinkers of the past have come to terms with these issues. Schrödinger clearly believes that there is more to the study of ancient history than mere factual curiosity and a concern with the origins of present-day thinking. His fascinatingly insightful study of the views of the philosopher/

scientists of antiquity, in *Nature and the Greeks*, makes clear that he also believes there is something directly to be gained from the Greeks' own insights, and what led them to their views, despite the undoubtedly enormous advances that modern science has made over what had been available to them at the time. Have we really made any progress at all concerning the really deep question: 'Whence come I and whither go I'? Schrödinger evidently believes not, though he appears to remain optimistic that genuine insights into such issues may become available to us in the future.

Having himself been one of the prime movers in the revolutionary changes that have taken place in our understanding of Nature at the scale of its tiniest ingredients, he is well placed to understand the importance of these changes in relation to what had been the views of physicists and philosophers immediately before him. Moreover, in my personal view, the more 'objective' philosophical standpoints of Schrödinger and Einstein with respect to quantum mechanics, are immeasurably superior to 'subjective' ones of Heisenberg and Bohr. While it is often held that the remarkable successes of quantum physics have led us to doubt the very existence of an 'objective reality' at the quantum level of molecules, atoms and their constituent particles, the extraordinary precision of the quantum formalism – which means, essentially, of the Schrödinger equation – signals to us that there must indeed be a 'reality' at the quantum level, albeit an unfamiliar one, in order that there can be a 'something' so accurately described by that very formalism.

Yet the formalism itself reveals a quantum-level reality that is strikingly different from the one that we experience at ordinary macroscopic scales. In a masterly way, Schrödinger paints for us a picture of that reality. I vividly recall, from my reading of *Science and Humanism* of forty years ago, Schrödinger's description of an iron letter-weight in the shape of a Great Dane that he had known as a small child, and that he retrieved after many years, having had to leave it behind in Austria when the Nazis came. What does it mean to say that it is the *same* dog as he had had before? There is no meaning to be attached to the 'sameness' of any of its individual particles. Schrödinger points out a remarkable irony. For over two thousand years, since the time of Leucippus and Democritus, there had been the fundamental idea that matter is composed of basic individual units, with empty space in between. Yet, this had been essentially a postulate, based on indirect inferences of widely differing acceptability. Then just as the first *direct* evidence of the atomistic nature of matter was beginning to come to light (such as in the Wilson cloud chamber and other experimental devices), quantum theory pulled the rug from beneath us. The particles that the theory revealed to us were not at all like the hard grains that we had come to expect, but were spread out in incomprehensible ways; worse still, they had no individuality whatever!

What is the present status of the particles that were known in Schrödinger's day? Electrons are still thought of as indivisible, but they belong to a larger family of particles, collectively called *leptons*. Protons, on the other hand, are not indivisible, being regarded as composed of still smaller units: the *quarks*. Modern particle physics is described in terms of these new kinds of element (quarks, leptons, gluons), which are the basic elements of what is referred to as the 'standard model'. In this model, the quarks and leptons are taken as structureless point-like objects. Are these the true atomic elements that physicists from the time of Leucippus and Democritus had sought?

I doubt that many present-day physicists would hold firmly to such a view. One prevalent line of thinking pins faith on the ideas of *string theory* according to which the basic units would not be point-like at all, but little loops referred to as 'strings'. These, however, would be far far tinier than the scales that are currently accessible to modern experimental techniques. There are some recent experimental indications that quarks may exhibit structure at much larger scales than those that would be required for string theory – in contradiction with the point-like expectations of the standard model. One must be cautious about drawing such conclusions, however, pending further results which may confirm or contradict them. This notwithstanding, it is fully to be expected that we are yet far from a final understanding of these matters.

In both of these books, Schrödinger shows himself to be deeply troubled, moreover, by the actual continuous nature of our pictures of space and time. According to quantum theory, the state of a material particle can undergo discontinuous jumps. In his attempts to reconcile this odd behaviour with the desirable feature that an individual particle ought really to retain some rudimentary sort of identity, Schrödinger is guided to the idea that it should be space itself, rather than the particles, which is discontinuous. I cannot help remarking, here, that this 'oddness' in the behaviour of quantum particles is now known to be even weirder than was imagined in Schrödinger's day. Schrödinger himself had pointed out, in 1935 (as a follow-up from some work by Einstein, Podolsky and Rosen), the puzzling phenomenon of *quantum entanglement*, according to which, in a system composed of more than one particle, the individual particles are not actually individual, but must be thought of as constituting an indivisible whole. In the mid-1960s John Bell showed that this entanglement could actually be directly measured, with consequences for our picture of reality that have still, in my opinion, not been adequately resolved.

Schrödinger, with considerable insight, goes back to ancient Greek times to try to examine the underlying reasons for our present firm beliefs in space-time continuity. He considers the picture of continuity that mathematicians, over the intervening centuries, have finally come to, and he points out the puzzling, almost paradoxical nature of this very picture. I had referred earlier to the powerful influence that Schrödinger had had on my own thinking. The idea that space and time are, at root, not what they 'seem' to be – perhaps themselves being discrete rather than continuous – is indeed something that took hold of me at that time, and the influence from Schrödinger's writings was great. I spent much time in trying to construct a theory in which spatial notions arose from an entirely discrete combinatorial structure. Although these attempts had some success, the thrust of underlying mathematical conceptions has been, instead, to drive us in the direction of that curiously elegant form of continuity that is provided by *complex numbers* (numbers in which $\sqrt{-1}$ features). Complex numbers are fundamental to quantum theory (and $\sqrt{-1}$ occurs explicitly in Schrödinger's equation). They are fundamental to the 'twistor theory' that my own deliberations led me to, and they are fundamental also to string theory. Moreover, they are fundamental to the deepest results of number theory (such as in Wiles's recent proof of Fermat's last theorem), which is the epitome of discrete mathematics. Perhaps, in complex numbers will ultimately be found the resolution between the discrete and continuous in physics that Schrödinger found so profoundly puzzling. Only time will tell.

Roger Penrose, March 1996

NATURE AND THE GREEKS

o

Shearman Lectures,
delivered at University College, London
on 24, 26, 28, and 31 May 1948

To my friend
A. B. CLERY
in gratitude for his
invaluable aid

CHAPTER I

THE MOTIVES FOR RETURNING TO ANCIENT THOUGHT

When, early in 1948, I set out to deliver a course of public lectures on the subject dealt with here, I still felt the urgent need of prefacing them with ample explanations and excuses. What I was expounding then and there (to wit, at University College, Dublin) has come to form a part of the little book before you. Some comment from the standpoint of modern science was added, and a brief exposition of what I deem to be the peculiar fundamental features of the present-day scientific world-picture. To prove that these features are historically produced (as against logically necessitated), by tracing them back to the earliest stage of Western philosophic thought, was my real objective in enlarging on the latter. Yet, as I said, I did feel a little uneasy, particularly since those lectures arose from my official duty as a professor of theoretical physics. There was need to explain (though I was myself not so thoroughly convinced of it) that in passing the time with narratives about ancient Greek thinkers and with comments on their views I was *not* just following a recently acquired hobby of mine; that it did not mean, from the professional point of view, a waste of time, which ought to be relegated to the hours of leisure; that it was justified by the hope of some gain in understanding modern science and thus *inter alia* also modern physics.

A few months later, in May, when speaking on the same topic at University College, London (Shearman Lectures, 1948), I already felt much more self-assured. While I had initially found myself supported mainly by such eminent scholars of antiquity as Theodor Gomperz, John Burnet, Cyril Bailey, Benjamin Farrington—some of whose pregnant remarks will later be quoted—I very soon became aware that it was probably neither haphazard nor personal predilection which made me plunge into the history of thought some twenty centuries deeper than other scientists had been induced to sound, who responded to the example and the exhortation of Ernst Mach. Far from following an odd impulse of my own, I had been swept along unwittingly, as happens so often, by a trend of thought rooted somehow in the intellectual situation of our time. Indeed, within the short period of one or two years several books had been published, whose authors were not classical scholars but were primarily interested in the scientific and philosophic thought of today; yet they had devoted a very substantial part of the scholarly labour embodied in their books to expounding and scrutinizing the earliest roots of modern thought in ancient writings. There is the posthumous *Growth of Physical Science* by the late Sir James Jeans, eminent astronomer and physicist, widely known to the public by his brilliant and successful popularizations. There is the marvellous *History of Western Philosophy* by Bertrand Russell, on whose manifold merits I need not and cannot enlarge here; I only wish to recall that Bertrand Russell entered his brilliant career as the philosopher of modern mathematics and mathematical logic. About one third of each of these volumes is concerned with antiquity. A handsome volume of a similar scope, entitled *The Birth of Science* (*Die Geburt der Wissenschaft*) was sent to me at nearly the same time from Innsbruck by the author, Anton von Mörl, who is neither a scholar of antiquity, nor of science, nor of philosophy; he had the misfortune at the time when Hitler marched into Austria to be the Chief of Police (*Sicherheitsdirector*) of Tirol, a crime for which he had to suffer many years in a concentration camp; he luckily survived the ordeal.

Now if I am right in calling this a general trend of our time, the questions naturally arise: how did it originate, what were its causes, and what does it really mean? Such questions can hardly ever be answered exhaustively even when the trend of thought that we consider lies far enough back in history for us to have gained a fair survey of the total human situation of the time. In dealing with a quite recent development one can at best hope to point out one or the other of the contributory facts or features. In the present case there are, I believe, two circumstances that may serve as a partial explanation of the strongly retrospective tendency among those concerned with the history of ideas: *one* refers to the intellectual and emotional phase mankind in general has entered in our days, the *other* is the inordinately critical situation in which nearly all the fundamental sciences find themselves ever more disconcertingly enveloped (as against their highly flourishing offspring like engineering, practical

—including nuclear—chemistry, medical and surgical art and technique). Let me briefly explain these two points, beginning with the first.

As Bertrand Russell has recently[1] pointed out with particular clarity, the growing antagonism between religion and science did not arise from accidental circumstances, nor is it, generally speaking, caused by ill will on either side. A considerable amount of mutual distrust is, alas, natural and understandable. One of the aims, if not perhaps the main task, of religious movements has always been to round off the ever unaccomplished understanding of the unsatisfactory and bewildering situation in which man finds himself in the world; to *close* the disconcerting 'openness' of the outlook gained from experience alone, in order to raise his confidence in life and strengthen his natural benevolence and sympathy towards his fellow creatures—innate properties, so I believe, but easily overpowered by personal mishaps and the pangs of misery. Now, in order to satisfy the ordinary, unlearned man, this rounding-off of the fragmentary and incoherent world picture has to furnish *inter alia* an explanation of all those traits of the material world that are either really not yet understood at the time or not in a way the ordinary unlearned man can grasp. This need is seldom overlooked for the simple reason that, as a rule, it is shared by the person or persons who, by their eminent characters, their sociable inclination, and their deeper insight into human affairs, have the power to prevail on the masses and to fill them with enthusiasm for their

[1] *Hist. West. Phil.* p. 559.

enlightened moral teaching. It so happens that such persons, as regards their upbringing and learning and apart from these extraordinary qualities, have usually themselves been quite ordinary men. Their views about the material universe would thus be as precarious, actually much the same, as those of their listeners. Anyhow, they would consider the spreading of the latest news about it irrelevant for their purpose, even if they knew them.

At first this mattered little or nothing. But in the course of the centuries, particularly after the rebirth of science in the seventeenth century, it came to matter a lot. According as, on the one hand, the teachings of religion were codified and petrified and, on the other hand, science came to transform—not to say disfigure—the life of the day beyond recognition and thereby to intrude into the mind of everyman, the mutual distrust between religion and science was bound to grow up. It did not spring from those well-known irrelevant details from which it ostensibly issued, such as whether the earth is in motion or at rest, or whether or not man is a late descendant of the animal kingdom; such bones of contention can be overcome, and to a large extent have been overcome. The misgiving is much more deeply rooted. By explaining more and more about the material structure of the world, and about how our environment and our bodily selves had, by natural causes, reached the state in which we find them, moreover by giving this knowledge away to everybody who was interested, the scientific outlook, so it was feared, stealthily wrested more and more from the hands of the Godhead, heading thus for a *self-contained* world to which God was in danger of becoming a gratuitous embellishment. It would hardly do justice to those who genuinely harboured this fear, if we declared it utterly unfounded. Socially and morally dangerous misgivings may spring, and occasionally have sprung—not, of course, from people knowing too much—but from people believing that they know a good deal more than they do.

Equally justified is, however, an apprehension which is, so to say, complementary and which has haunted science from the very time it came into existence. Science has to be careful of incompetent interference from the other side, particularly in scientific disguise, recalling Mephisto, who, in the borrowed robe of the Doctor, foists his irreverent jokes upon the ingenuous Scholar. What I mean is this. In an honest search for knowledge you quite often have to abide by ignorance for an indefinite period. Instead of filling a gap by guesswork, genuine science prefers to put up with it; and this, not so much from conscientious scruples about telling lies, as from the consideration that, however irksome the gap may be, its obliteration by a fake removes the urge to seek after a tenable answer. So efficiently may attention be diverted that the answer is missed even when, by good luck, it comes close at hand. The steadfastness in standing up to a *non liquet*, nay in appreciating it as a stimulus and a signpost to further quest, is a natural and indispensable disposition in the mind of a scientist. This in itself is apt to set him at variance with the religious aim of closing the picture, unless each of the two antagonistic attitudes, both legitimate for their respective purposes, is applied with prudence.

Such gaps easily evoke the impression of being undefended weak spots. They are at times seized upon by persons whom they please, not as an incentive for further quest, but as an antidote against their fear that science might, by 'explaining everything', deprive the world of its metaphysical interest. A new hypothesis is put up, as everybody is, of course, entitled to do in such a case. At first sight it seems firmly anchored in obvious facts; one only wonders why these facts or the ease with which the proposed explanation follows from them have escaped everybody else. But this in itself is no objection, for it is precisely the situation we very often have to face in the case of genuine discoveries. However, on closer inspection the enterprise betrays its character (in the cases I have in mind) by the fact that, while apparently tendering an acceptable explanation within a fairly wide range of inquiry, it is at variance with generally established principles of sound science, which it either pretends to overlook or airily reduces with regard to their generality; to believe in the latter, so we are told, was just the prejudice that was in the way of a correct interpretation of the phenomena in question. But the creative vigour of a general principle depends precisely on its generality. By losing ground it loses all its strength and can no longer serve as a reliable guide, because in every single instance of application its competence may be challenged. To clinch the suspicion that this dethronement was not an accidental by-product of the whole enterprise, but its sinister goal, the territory from which previous scientific attainment is invited to retire is with admirable dexterity claimed as a playground of some religious ideology that cannot really use it profitably, because its true domain is far beyond anything in reach of scientific explanation.

A well-known instance of such intrusion is the recurring attempt to reintroduce *finality* into science, allegedly because the reiterated crises of *causality* prove it to be incompetent single-handed, actually because it is considered *infra dig.* of God Almighty to create a world which He disallowed Himself to tamper with ever after. In this case the weak spots seized upon are obvious. Neither in the theory of evolution nor in the mind-matter problem has science been able to adumbrate the causal linkage satisfactorily even to its most ardent disciples. And so *vis viva, élan vital,* entelechy, wholeness, directed mutations, quantum mechanics of free will, etc. stepped in. As a curiosity, let me mention a neat volume[1] printed on much better paper and in much more handsome form than British authors were used to at that time. After a sound and scholarly report on modern physics, the author happily embarks on the teleology, the purposiveness, of the interior of the atom and interprets in this manner all its activities, the movements of the electrons, the emission and absorption of radiation, etc.,

> And hopes to please by this peculiar whim
> The God who fashioned it and gave it him.[2]

[1] Zeno Bucher, *Die Innenwelt der Atome* (Lucerne: Josef Stocker, 1946).

[2] From Kenneth Hare, *The Puritan.*

But let me return to our general topic. I was trying to set forth the intrinsic causes for the natural enmity between science and religion. The fights that sprang from it in the past are too well known to call for further comment. Moreover, they are not what concerns us here. However deplorable, they still manifested mutual interest. Scientists on the one side, and metaphysicians, both of the official and of the learned type, on the other, were still aware that their endeavours to secure insight referred after all to the same object—man and his world. A clearance of the widely diverging opinions was still felt a necessity. It has not been attained. The comparative truce we witness today, at least among cultured people, was not reached by setting in harmony with one another the two kinds of outlook, the strictly scientific and the metaphysical, but rather by a resolve to ignore each other, little short of contempt. In a treatise on physics or biology, albeit a popular one, to digress to the metaphysical aspect of the subject is considered impertinent, and if a scientist dare, he is liable to have his fingers rapped and be left to guess whether it is for offending science or the particular brand of metaphysics to which the critic is devoted. It is pathetically amusing to observe how on the one side only scientific information is taken seriously, while the other side ranges science among man's worldly activities, whose findings are less momentous and have, as a matter of course, to give way when at variance with the superior insight gained in a different fashion, by pure thought or by revelation. One regrets to see mankind strive towards the same goal along two different and difficult winding paths, with blinkers and separating walls, and with little attempt to join all forces and to achieve, if not a full understanding of nature and the human situation, at least the soothing recognition of the intrinsic unity of our search. This is regrettable, I say, and would be a sad spectacle anyhow, because it obviously reduces the range of what could be attained if all the thinking power at our disposal were pooled without bias. However, the loss might perhaps be endured if the metaphor I used were really appropriate, that is to say, if it were actually two different crowds who follow two paths. But this is not so. Many of us are not decided which one to follow. With regret, nay with despair, many find that they have to shut themselves off alternately from the one and from the other kind of outlook. It is certainly not in general the case that by acquiring a good all-round scientific education you so completely satisfy the innate longing for a religious or philosophical stabilization, in face of the vicissitudes of everyday life, as to feel quite happy without anything more. What does happen often is that science suffices to jeopardize popular religious convictions, but not to replace them by anything else. This produces the grotesque phenomenon of scientifically trained, highly competent minds with an unbelievably childlike—undeveloped or atrophied—philosophical outlook.

If you live in fairly comfortable and secure conditions, and take them to be human life's general pattern, which, thanks to inevitable progress, wherein you believe, is about to spread and to become universal, you seem to get along quite well without any philosophical outlook; if not indefinitely, at least until you grow old and decrepit and begin to face death as a reality. But while the early stages of the rapid material advancement which came in the wake of modern science appeared to inaugurate an era of peace, security and progress, this state of affairs now no longer prevails. Matters have sadly changed. Many people, indeed entire populations, have been thrown out of their comfort and safety, have suffered inordinate bereavements, and look into a dim future for themselves and for those of their children who have not perished. The very survival, let alone the continued progress, of man is no longer regarded as certain. Personal misery, buried hopes, impending disaster, and distrust of the prudence and honesty of the wordly rulers are apt to make men crave for even a vague hope, whether rigorously provable or not, that the 'world' or 'life' of experience be embedded in a context of higher, if as yet inscrutable, significance. But there is the wall, separating the 'two paths', that of the heart and that of pure reason. We look back along the wall: could we not pull it down, has it always been there? As we scan its windings over hills and vales back in history we behold a land far, far, away at a space of over two thousand years back, where the wall flattens and disappears and the path was not yet split, but was only *one*. Some of us deem it worth while to walk back and see what can be learnt from the alluring primeval unity.

Dropping the metaphor, it is my opinion that the philosophy of the ancient Greeks attracts us at this moment, because never before or since, anywhere in the world, has anything like their highly advanced and articulated system of knowledge and speculation been established *without* the fateful division which has hampered us for centuries and has become unendurable in our days. There were, of course, widely diverging opinions, combating one another with no less fervour, and occasionally with no more honourable means—such as unacknowledged borrowing and destruction of writings—than elsewhere and at other periods. But there was no limitation as to the subjects on which a learned man would be allowed by other learned men to give his opinion. It was still agreed that the true subject was essentially one, and that important conclusions reached about any part of it could, and as a rule would, bear on almost every other part. The idea of delimitation in water-tight compartments had not yet sprung up. A man could easily find himself blamed, conversely, for shutting his eyes to such interconnexion—as were the early atomists for being silent on the consequences in ethics, of the universal necessity which they assumed and for failing to explain how the motions of the atoms and those observed in the skies had originally been set up. To put it dramatically: one can imagine a scholar of the young School of Athens paying a holiday visit to Abdera (with due caution to keep it secret from his Master), and on being received by the wise, far-travelled and world-famous old gentleman Democritus, asking him questions on the atoms, on the shape of the earth, on moral conduct, God, and the immortality of the soul—without being repudiated on any of these points. Can you easily imagine such a motley conversation between a student and his teacher in our days? Yet, in all probability, quite a few young people have a similar—we should say quaint—collection of inquiries on their minds, and would like to discuss all of them with the one person of their confidence.

So much for the first of the two points that I announced my intention of submitting as clues to the renascent interest in ancient thought. Let me now put forward the second point, namely, the present crisis of the fundamental sciences.

Most of us believe that an ideally accomplished science of the happenings in space and time would be able to reduce them in principle to events that are completely accessible and understandable to (an ideally accomplished) physics. But it was from physics that, in the early years of the century, the first shocks—quantum theory and the theory of relativity—started to set the foundations of science trembling. During the great classical period of the nineteenth century, however remote might seem the realization of the task of actually describing in terms of physics the growth of a plant or the physiological processes in the brain of a human thinker or of a swallow building its nest, the language in which the account ought eventually to be drawn up was believed to be deciphered, namely: corpuscles, the ultimate constituents of matter, move under their mutual interaction, which is not instantaneous, but transmitted by a ubiquitous medium that one may or may not choose to call ether;

the very terms 'movement' and 'transmission' imply that the measure and the scene of all this are time and space; these have no other property or task than to be the stage, as it were, on which we image the corpuscles moving and their interaction being transmitted. Now, on the one hand, the relativistic theory of gravitation goes to show that the distinction between 'actor' and 'stage' is not expedient. Matter and the (field- or wave-like) propagation of something transmitting the interaction ought better to be regarded as the *shape* of space-time itself, which ought not to be looked upon as being conceptually prior to what was hitherto called its content; no more than, say, the corners of a triangle are prior to the triangle. Quantum theory, on the other hand, tells us that what was formerly considered as the most obvious and fundamental property of the corpuscles, so much so that it was hardly ever mentioned, namely their being identifiable individuals, has only a limited significance. Only when a corpuscle is moving with sufficient speed in a region not too crowded with corpuscles of the same kind does its identity remain (nearly) unambiguous. Otherwise it becomes blurred. And by this assertion we do not mean to indicate merely our practical inability to follow the movement of the particle in question; the very notion of absolute identity is believed to be inadmissible. At the same time we are told that the interaction, whenever it has —as it frequently has—the form of waves of short wave-length and low intensity, itself assumes the form of fairly well identifiable particles—in the teeth of the aforesaid description as waves. The particles which represent the interaction in the course of its propagation are, in every particular case, different in kind from those that interact; yet they have the same claim to be called particles. To round off the picture, particles of any kind exhibit the character of waves which becomes the more pronounced the slower they move and the denser they crowd, with the corresponding loss of individuality.

The argument for whose sake I have inserted this brief report would be reinforced by mentioning the 'pulling down of the frontier between observer and observed' which many consider an even more momentous revolution of thought, while to my mind it seems a much overrated provisional aspect without profound significance. Anyhow, my point is this. The modern development, which those who have brought it to the fore are yet far from really understanding, has intruded into the relatively simple scheme of physics which towards the end of the nineteenth century looked fairly stabilized. This intrusion has, in a way, overthrown what had been built on the foundations laid in the seventeenth century, mainly by Galileo, Huygens and Newton. The very foundations were shaken. Not that we are not everywhere still under the spell of this great period. We are all the time using its basic conceptions, though in a form their authors would hardly recognize. And at the same time we are aware that we are at the end of our tether. It is, then, natural to recall that the thinkers who started to mould modern science did not begin from scratch. Though they had little to borrow from the earlier centuries of our era, they very truly revived and continued ancient science and philosophy. From this source, awe-inspiring both by its remoteness in time and by its genuine grandeur, pre-conceived ideas and unwarranted assumptions may have been taken over by the fathers of modern science, and would, by their authority, soon be perpetuated. Had the highly flexible and open-minded spirit that pervaded antiquity continued, such points would have continued to be debated and could have been corrected. A prejudice is more easily detected in the primitive, ingenuous form in which it first arises than as the sophisticated, ossified dogma it is apt to become later. Science does appear to be baffled by ingrained habits of thought, some of which seem to be very difficult to find out, while others have already been discovered. The theory of relativity has done away with Newton's concepts of absolute space and time, in other words of absolute motionlessness and absolute simultaneity, and it has ousted the time-honoured couple 'force and matter' at least from its dominating position. Quantum theory, while extending atomism almost limitlessly, has at the same time plunged it into a crisis that is severer than most people are prepared to admit. On the whole the present crisis in modern basic science points to the necessity of revising its foundations down to very early layers.

This, then, is a further incentive for us to return once again to an assiduous study of Greek thought. There is not only, as was pointed out earlier in this chapter, the hope of unearthing obliterated wisdom, but also of discovering inveterate error at the source, where it is easier to recognize. By the serious attempt to put ourselves back into the intellectual situation of the ancient thinkers, far less experienced as regards the actual behaviour of nature, but also very often much less biased, we may regain from them their freedom of thought—albeit possibly in order to use it, aided by our superior knowledge of facts, for correcting early mistakes of theirs that may still be baffling us.

Let me conclude this chapter by some quotations. The first bears closely on what has just been said. It is translated from Theodor Gomperz's *Griechische Denker*.[1] To meet the possible objection that no practical advantage can spring from studying ancient opinion, which has been long superseded by better insight based on vastly superior information, a series of arguments is brought to the fore that ends with the following notable paragraph:

> It is of even greater importance to recall an *indirect* kind of application or utilization that must be regarded as highly momentous. Nearly our entire intellectual education originates from the Greeks. A thorough knowledge of these origins is the indispensable prerequisite for *freeing* ourselves from their overwhelming influence. To ignore the past is here not merely undesirable, but simply impossible. You need not know of the doctrines and writings of the great masters of antiquity, of Plato and Aristotle, you need never have heard their names, none the less you are under the spell of their authority. Not only has their influence been passed on by those who took over from them in ancient and in modern times; our entire thinking, the logical categories in which it moves, the linguistic patterns it uses (being therefore dominated by them)—all this is in no small degree an artefact and is, in the main, the product

[1] Vol. I, p. 419 (3rd ed. 1911).

of the great thinkers of antiquity. We must, indeed, investigate this process of becoming in all thoroughness, lest we mistake for primitive what is the result of growth and development, and for natural what is actually artificial.

The following lines are taken from the Preface of John Burnet's *Early Greek Philosophy*: '...it is an adequate description of science to say that it is "thinking about the world in the Greek way". That is why science has never existed except among peoples who came under the influence of Greece.' This is the most concise justification a scientist could wish for, to excuse his propensity for 'wasting his time' in studies of this kind.

And an excuse seems to be needed. Ernst Mach, the physicist colleague of Gomperz at the University of Vienna, and eminent historian (!) of physics, had, a few decades earlier, spoken of the 'scarce and poor remnants of ancient science'.[1] He continues thus:

> Our culture has gradually acquired full independence, soaring far above that of antiquity. It is following an entirely *new* trend. It centres around mathematical and scientific enlightenment. The traces of ancient ideas, still lingering in philosophy, jurisprudence, art and science constitute impediments rather than assets, and will come to be untenable in the long run in face of the development of our own views.

For all its supercilious crudeness, Mach's view has a relevant point in common with what I have quoted from Gomperz, namely the plea for our having to *overcome* the Greeks. But while Gomperz supports a non-trivial turn by obviously true arguments, Mach

[1] *Popular Lectures*, 3rd ed., essay no. XVII (J. A. Barth, 1903).

clinches the trivial side by gross exaggeration. In other passages of the same paper he recommends a quaint method of getting beyond antiquity, namely to neglect and ignore it. In this, for all I know, he had little success—fortunately, for the mistakes of the great, promulgated along with the discoveries of their genius, are apt to work serious havoc.

CHAPTER II

THE COMPETITION, REASON *v.* SENSES

The short passage from Burnet and the longer one quoted from Gomperz at the end of the last chapter form the selected 'text', as it were, of this little book. We shall return to them later, when we shall try to answer the question: what is, then, that Greek way of thinking about the world? What are those peculiar traits, in our present scientific world view, that originated from the Greeks, whose special inventions they were, that are thus not necessary but artificial, being only historically produced and thus capable of change or modification, and which we, by ingrained habit, are liable to regard as natural and inalienable, as the only possible way of looking at the world?

However, at the moment we shall not yet enter on this main question. Rather, by way of preparing the answer, I wish to introduce the reader to parts of ancient Greek thought which I consider relevant in our context. In this I shall not adopt a chronological arrangement. For I am neither willing nor competent to write a brief history of Greek philosophy, there being so many good, modern and attractive ones (particularly Bertrand Russell's and Benjamin Farrington's) at the disposal of the reader. Instead of following the order in time let us be guided by the intrinsic connexion of the subjects. This will bring together various thinkers' ideas on the same problem rather than the attitude of a single philosopher, or of a group of sages, towards the most various questions. It is the ideas we wish to reconstruct here, not the separate persons or minds. So we shall choose two or three leading ideas or motives of thought, which arose at an early stage, kept minds alert during the centuries of antiquity, and are in intimate connexion, if not identical, with problems enjoying the full vigour of agitated dispute up to the present moment. Grouping the tenets of ancient thinkers around these leading ideas, we shall feel their intellectual joys and grievances nearer to our own than is sometimes suspected.

A widely discussed question, given great prominence in the natural philosophy of the ancients from the very beginning down through the centuries, deals with the reliability of the senses. This, anyhow, is the heading under which the problem is often reviewed in modern scholarly treatises. It arose from observing that the senses occasionally 'deceive' us—as when a straight bar, half immersed obliquely in water, appears broken—and from noticing that the same object affects different persons differently—the current instance in antiquity being honey tasting bitter to the jaundiced. Until not very long ago some scientists used to be content with the distinction between what they chose to call the 'secondary' qualities of matter, colour, taste, smell, etc., and its 'primary' qualities, extension and motion. This distinction was no doubt a late descendent of the old controversy, an attempted settlement: the primary qualities were thought to be the extract, the true and unshakable, distilled by reason from the direct yield of our sense data. This view is, of course, no longer acceptable, since we have learnt from the theory of relativity (if we did not know it before) that space and time, and the shape and motion of matter in space and time, are an elaborate hypothetical construction of the mind, not at all unshakable, much less so than the direct sensates, which, if anything, deserve the epithet 'primary'.

But the reliability of the senses is only the preamble to much deeper questions, which are very much alive today and of which some of the ancient thinkers were fully aware. Is our attempted world picture based on sense perceptions alone? What role does reason play in its construction? Does it perhaps ultimately and truly rest upon pure reason alone?

Amid the triumphant march of experimental discoveries in the nineteenth century any philosophical view with a strong leaning towards 'pure reason' received very bad marks, certainly from the leading scientists. This is no longer so. The late Sir Arthur Eddington became more and more affectionately attached to the pure reason theory. Though few would follow him to the extreme, his exposition was admired for being ingenious and fruitful. Max Born found it necessary after all to write a pamphlet in refutation. Sir Edmund Whittaker was, to say the least, very much taken with Eddington's claim that some ostensibly purely empirical constants can be inferred from pure reason, for instance, the total number of elementary particles in the universe. Disregarding details and taking a wider view of Eddington's endeavour, which sprung from a strong confidence in the reasonableness and simplicity of nature, we find his ideas by no means isolated. Even Einstein's marvellous theory of gravitation, based on sound experimental evidence and firmly clinched by new observational facts which he had predicted, could only be discovered by a genius with a strong feeling for the simplicity and beauty of ideas. The attempts to generalize his great successful conception, so as to embrace electromagnetism and the interaction of nuclear particles, are informed by the hope of 'guessing' to a large measure the way in which nature really works, of getting the clue from the principle of simplicity and beauty. In fact traces of this attitude pervade the work in modern theoretical physics—maybe too much so, but this is not the place for criticism.

The extreme viewpoints in respect of the attempt to construct *a priori* from reason the actual behaviour of nature can be represented in recent times by the names of Eddington at the one end and of, let me say, Ernst Mach at the other. The full range of possible attitudes within these limits and the full vigour of clinging to one view, defending it and attacking, nay ridiculing, the rejected alternative has notable representatives among the great thinkers of antiquity. We really do not know whether we should be astonished that they, with their infinitely inferior knowledge of the actual laws of nature, could have developed all the diversity of opinion about their foundation and the hot-headed zeal in defending the personally favoured one, or should rather wonder that the controversy has not yet subsided, quenched by the far-reaching insight we have gained since.

Parmenides, who flourished in Elea, Italy, around 480 B.C. (which is roughly a decade before Socrates was born in Athens, and a little more than a decade before the birth of Democritus in Abdera), is one of the first to develop an extremely anti-sensual, aprioristically conceived view of the world. His world contained very little, so little in fact and that little in so flat contradiction of observed facts that he was induced to give, along with his 'true' conception of it, an attractive description of (as we should say) 'the world as it really is', with sky, sun, moon and stars and certainly many other things. But this, he said, was only our belief, it was all due to the deception of the senses. In truth there were not many things in the world but only One Thing. And this One Thing is (I beg your pardon) the thing that *is*, in contradistinction to the thing that is not. This latter, from pure logic, is *not*—and thus there *is* only the One Thing, first mentioned. Moreover, there can be no place in space nor any moment in time where or when the One is not—for being the thing that is, it can nowhere and never have the contradictory predication that it be not. Hence it is ubiquitous and eternal. There can be no change and no movement, since there is no empty space into which the One could move where it is not already. All we believe to witness to the contrary is deception.

The reader will notice that we are faced with a religion—recited, by the way, in fine Greek verse—rather than with a scientific world-view. But at the time such a distinction would not appear. Religion or piety towards the gods would, for Parmenides, doubtless belong to the apparent world of 'belief'. His 'truth' was the purest monism ever conceived. He became the father of a school (the Eleatics) and had an enormous influence on the generations to come. Plato took the Eleatic objections to his 'theory of forms' very seriously. In the dialogue he named after our sage and dated back before his own birth (to the time when Socrates was a young man) Plato expounds these objections but hardly attempts to refute them.

Let me fill in a detail which is perhaps more than a detail. From my above brief characterization, in which I followed the usual rendering, it might seem that Parmenides' dogmatism referred to the material world, which he replaced by something else according to his liking and in flat contradiction to observation. But his monism was deeper than that. In one of the texts quoted by Diels[1], Parmenides fragment 5

> "for the same is the thinking and the being"

follows immediately (with an implication of similarity of meaning) on a quotation from Aristophanes "thinking has the same power as doing". Again in the first line of fr. 6 we read:

> both the saying and the thinking is a thing that *is*;

and in fr. 8, lines 34 f.,

> One and the same is the thinking and that for whose sake the thought is there.

(I have followed Diels' interpretation and waived Burnet's objection, that the definite article would be

[1] Diels, *Die Fragmente der Vorsokratiker* (Berlin, 1903), 1st ed.

required to make the Greek infinitives which I have rendered by 'the thinking' and 'the being' the subjects of the sentence. In Burnet's translation fr. 5 loses the akinness to Aristophanes' statement, while the line from fr. 8 becomes flatly tautological in Burnet's rendering: 'The thing that can be thought and that for the sake of which the thought exists is the same.')

Let me add a remark of Plotinus (quoted by Diels for fr. 5) in which he says that Parmenides 'united into one the thing that is and reason, and would not put the thing that is in the sensates. Saying "for the same is the thinking and the being", he also says that the latter is motionless, even though in joining the thinking he deprives it of all bodylike motion.' [...*εὶς ταὐτὸ συνῆγεν ὂν καὶ νοῦν καὶ τὸ ὂν οὐκ ἐν τοῖς αἰσθητοῖς ἐτίθετο. 'τὸ γὰρ αὐτὸ νοεῖν ἐστίν τε καὶ εἶναι' λέγων καὶ ἀκίνητον λέγει τοῦτο, καίτοι προστιθεὶς τὸ νοεῖν σωματικὴν πᾶσαν κίνησιν ἐξαιρῶν ἀπ' αὐτοῦ.*]

From this repeated emphasis on the identity of the *ὄν* (the thing that is) and the *νοεῖν* (thinking) or *νόημα* (thought), and from the way his assertions were referred to by thinkers of antiquity, we must infer that Parmenides' motionless, eternal One was not meant to be a whimsical, distorted and inadequate mental image of the real world around us, as if its true nature were that of a homogeneous, unstirring fluid, filling for ever the whole of space without boundaries—a simplified, hyperspherical Einstein-universe, as the modern physicist would be inclined to call it. His attitude is that he does not take the material world around us as a granted reality. The true reality he puts into thought, into the subject of cognizance as we should say. The world around us is a product of the sensates, an image created by the sense perception in the thinking subject 'by way of belief'. That he deems it well worth consideration and description, the poet-philosopher shows by the second half of his poem, which is entirely devoted to it. But what the senses yield to us is not the world as it really is, not the 'thing in itself' as Kant put it. The latter resides in the subject, in the fact that it is a subject, capable of thinking, capable of some mental process at least—of permanently willing, as Schopenhauer had it. I have no doubt that this is our philosopher's eternal, motionless One. It remains intrinsically unaffected, unchanged by the passing show that the senses display to it—the same as Schopenhauer asserted of the Will, which he tried to explain was Kant's 'thing in itself'. We are faced with a poetical attempt—poetical not only as regards the metrical form—at a union of the Mind (or if you like the Soul), the World and the Godhead. Confronted with the intensely perceived oneness and changelessness of the Mind, the apparent kaleidoscopic character of the World had to give way and be regarded as a mere illusion. Clearly this results in an impossible distortion, which is remedied, as it were, by the second part of Parmenides' poem.

It is true that this second part implies a grave inconsistency, which could not, however, be removed by any interpretation. If reality is abrogated from the material world of the senses, is the latter then a μὴ ὄν, a thing that actually does *not* exist? And is then the second part a fairy tale, all about things that are not? But, at least, it is said to deal with human *beliefs* (δόξαι); they are in the mind (νοεῖν), which is identified with existence (εἶναι); have they then not a certain existence as phenomena of the mind? These are questions we cannot answer, contradictions we cannot remove. We must be content to remember that he who touches for the first time at a deep, hidden truth that is contrary to universally accepted opinion usually overstates it in a way that is likely to involve him in logical contradictions.

We now turn to consider briefly the views of someone who represents the other extreme in the scale of possible attitudes towards the question, whether the direct sensual information or the reasoning human mind is the chief source of truth and has thus the fuller, or even the only claim to reality, properly speaking. As an outstanding example of pure sensualism we adduce the great sophist Protagoras, who was born around 492 B.C. in Abdera (which a generation later, around 460 B.C., was to give birth to the great Democritus). Protagoras regarded the sense perceptions as the only things that really existed, the only material from which our world-picture is made up. In principle all of them have to pass for equally true, even when modified or distorted by fever, disease, intoxication or madness. The stock example in antiquity was the bitter taste that honey has for the jaundiced, while to other persons it seemed sweet. Protagoras would have nothing of 'seeming' or illusion in either case, though it was, he said, our duty to try and cure people possessed of similar anomalies. He was not a scientist (any more than Parmenides was), though he did have profound interest in the Ionian enlightenment (of which we shall have to speak later). According to B. Farrington the efforts of Protagoras were centred upon standing up for human rights in general, upon promoting a more equitable social system, equal citizen rights for all human beings—true democracy, in short. In this, of course, he did not succeed, since ancient culture continued till its downfall to rest upon an economic and social system which depended vitally on the *in*equality of human beings. His best known saying, that 'man is the measure of all things', is usually taken to refer to his sensual theory of knowledge, but might well embrace a plainly human attitude towards the political and social question: human affairs to be ordered by laws and customs suited to the nature of man, and unprejudiced by tradition or superstition of any kind. His attitude to traditional religion is preserved in the following words which are as cautious as they are witty: 'With regard to the gods, I cannot know either that they are or that they are not, or what they are like in figure, for there are many things that hinder sure knowledge, the obscurity of the subject and the shortness of human life.'

The most advanced epistemological attitude I encountered in any thinker of antiquity is clearly and pregnantly expressed in at least one of the fragments of Democritus. We shall have to come back to him as the great atomist. Suffice it for the moment to say that he certainly believed in the expediency of the material world-view to which he had been led, believed it as firmly as any physicist of our times: the rigid, immutable little corpuscles that move in empty space along straight lines, collide, rebound, etc., etc. and thus produce all the immense variety of what is observed in the material world. He believed in this reduction of the unspeakably rich manifold of goings-on to purely geometrical images, and he was right in his belief. Theoretical physics was at that time so far ahead of experiment (which was hardly known) as never before or after—not to speak of our own days, which see it scramble along in the rear. Yet Democritus at the same time realized that the naked intellectual construction which in his world-picture had supplanted the actual world of light and colour, sound and fragrance, sweetness, bitterness and beauty, was actually based on nothing but the sense perceptions themselves which had ostensibly vanished from it. In fragment D 125, taken from Galen and discovered only about fifty years ago, he introduces the intellect (*διάνοια*) in a contest with the senses (*αἰσθήσεις*). The former says: 'Ostensibly there is colour, ostensibly sweetness, ostensibly bitterness, actually only atoms and the void'; to which the senses retort: 'Poor intellect, do you hope to defeat us while from us you borrow your evidence? Your victory is your defeat.' You simply cannot put it more briefly and clearly.

Numerous other fragments of this great thinker might be typical places from Kant's work: that we cognize nothing as it really is, that we truly know nothing, truth is hidden deep in the dark, and so on.

Scepticism alone is a cheap and barren affair. Scepticism in a man who has come nearer to the truth than anyone before, and yet clearly recognizes the narrow limits of his own mental construction, is great and fruitful, and does not reduce but doubles the value of the discoveries.

CHAPTER III

THE PYTHAGOREANS

From men like Parmenides or Protagoras we can infer little or nothing as to the scientific efficacy of such extreme viewpoints as they held, for they were not scientists. The prototype of a school of thinkers with strongly scientific orientation and at the same time with a well-marked bias, bordering on religious prejudice, towards reducing the edifice of nature to pure reason were the Pythagoreans. Their main seat was southern Italy, the towns of Croton, Sybaris, Tarentum around the bay between the 'heel' and the 'toe' of the peninsula. The adherents formed something very much like a religious order with quaint rites as to food and other things, bound to secrecy towards outsiders, at least on parts of the teaching.[1] The founder, Pythagoras, who flourished in the second half of the sixth century B.C., must have been one of the most remarkable persons of antiquity, around whom legends of supernatural power sprouted: that he could remember all the previous lives in his metempsychoses (migration

[1] Various ancient authors comment on a great scandal which Hippasus caused by divulging the existence of the pentagon-dodecahedron, or, as others say, a certain 'incommensurability' (ἀλογία) and 'asymmetry'. He was expelled from the Order. Other punishments are mentioned: his grave was prepared for him as for a defunct; he was (by the avenging godhead) drowned on the high sea.

Another big scandal in antiquity is connected with the rumour that Plato purchased at a high price, from a Pythagorean who was in need of money, three manuscript rolls, in order to use them for himself without divulging his sources.

of the soul); that someone on an accidental shifting of his garment had noticed that his thigh was of pure gold. He seems to have left not a line in writing. His word was gospel to his pupils, as is evidenced by the well-known *αὐτὸς ἔφα* ('the Master said so'), which would settle any dispute among them and clinch the infallible truth. It is also said that they were awed to pronounce his name but spoke of him as 'yonder man' (*ἐκεῖνος ἀνήρ*). But it is sometimes not easy for us to decide whether a particular doctrine goes back to him, or from whom it originated, on account of the above-mentioned character and attitude of the community.

Their aprioristic outlook was visibly taken over by Plato and the Academy, who were deeply impressed and strongly influenced by the South Italian school. Indeed, from the point of view of the history of ideas, we might very well call the Athenian school a branch of the Pythagoreans. That they did not formally adhere to the 'Order' is of little relevance, and of still less is their anxiety to veil rather than to emphasize their dependence with a view to enhancing their own originality. But our best information on the Pythagoreans we owe, as we owe so much other information, to the sincere and honest reports of Aristotle, even though he mostly disagrees with their views and blames them for unfounded aprioristic bias, to which he himself was so liable.

The basic doctrine of the Pythagoreans, we are told, was that *things are numbers*, though some reports try to weaken the paradox, saying 'are like numbers', analogous to numbers. We are far from knowing what was really meant by this assertion. It very likely originated, as a sweeping generalization of truly imposing boldness and grandeur, from Pythagoras' famous discovery of the integral or rational subdivisions (for instance $\frac{1}{2}$, $\frac{2}{3}$, $\frac{3}{4}$) of a string, producing musical intervals which, when composed in the harmony of a song, may move us to tears, speaking as it were, directly to the soul. (A beautiful simile of the relation between soul and body originated in the School, probably from Philolaus: the soul is called the harmony of the body, related to it as are to a musical instrument the sounds it produces.)

According to Aristotle the 'things' (that were numbers) were in the first place sensual, material objects; for instance, after Empedocles had developed his theory of the four elements, they too 'became' numbers; but also such 'things' as Soul, Justice, Opportunity had, or 'were', their numbers. In the allotment some simple properties of number-theory were relevant. For instance, square numbers (4, 9, 16, 25, ...) had to do with Justice, which was more particularly identified with the first of them, namely with 4. Here the underlying idea must have been the possibility of splitting the number into two *equal* factors (compare words like 'equity', 'equitable'). A square number of dots can be arranged in a square, as for example in ninepins. In the same way the Pythagoreans spoke of triangular numbers, such as 3, 6, 10, ...

The number is obtained by multiplying the number of dots along one side (n) by the following one ($n+1$) and dividing the product (which is always even) by two, thus $\frac{n(n+1)}{2}$. (This is most easily seen by juxtaposing a second triangle upside down and shifting the figure to form a rectangle.

In modern theory the 'square of the orbital moment of momentum' is $n(n+1)h^2$, not n^2h^2, n being an integer. This remark is only to illustrate the fact that the distinction of triangular numbers was not a mere illusion, they do quite often make their appearance in mathematics.)

The triangular number 10 enjoyed singular respect, possibly because it was the fourth, and thus the one pointing to justice.

The amount of arrant nonsense that is bound to be produced along such lines we illustrate from Aristotle's faithful—and *not* sneering—report. The primary property of a number is the Odd or the Even. (So far so good. The mathematician is familiar with the fundamental distinction between odd and even *prime* numbers, even though the latter class contains only the one number 2.) But then the Odd is supposed to determine the limited or finite character of a thing, the Even is made responsible for the unlimited or infinite character of some things. It symbolizes infinite (!)

divisibility, because an even number can be divided into two equal parts. Another commentator finds a defectiveness or incompleteness (pointing to the infinite) of the even number in the fact that when you divide it in two

. . . | . . .

an empty field remains in the middle that has no master and no number (*ἀδέσποτος καὶ ἀνάριθμος*).

The four elements (fire, water, earth, air) seem to have been thought of as built up of four of the five regular bodies, while the fifth, the dodecahedron, was reserved as a receptacle for the whole universe, probably because it was so near to a sphere and was bounded by pentagons; this figure itself played a mystical role and so did the figure enhanced by its five diagonals ($5 + 5 = 10$) which form the well-known pentagram. One of the early Pythagoreans, Petron, contended that there were altogether 183 worlds, arranged in a triangle—though, by the way, this is not a triangular number. Is it very irreverent to remember on this occasion that we were recently told by an eminent scientist that the total number of elementary particles in the world was $16 \times 17 \times 2^{256}$, where 256 is the square of the square of the square of 2?

The later Pythagoreans believed in the transmigration of the soul in a very literal sense. It is usually said that Pythagoras himself did. Xenophanes, in a couple of distichs, tells us this anecdote about the master: when he passed a little dog which was being cruelly beaten, he was touched by pity and addressed the tormentor thus: 'Stop beating him; for it is the soul of a friend, whom I recognized on hearing his voice.' This on the part of Xenophanes was probably meant to ridicule the great man for his foolish belief. We cannot help feeling differently about it today. Supposing the story to be true, one might surmise a much simpler meaning for his words, just this: Stop, for I hear the voice of a tormented friend, calling for my help. ('Our friend the dog' became a standing phrase with Charles Sherrington.)

Let me return for a moment to the general idea, mentioned at the outset, the idea that numbers are at the back of everything. I said that it obviously started from the acoustical discoveries about the lengths of vibrating strings. But to do it justice (in spite of its crazy offshoots) one must not forget that this was the time and the place of the first great discoveries in mathematics and geometry, which were usually connected with some actual or imagined application to material objects. Now the essence of mathematical thought is that from the material setting it abstracts numbers (lengths, angles and other quantities) and deals with them and their relations as such. It is in the nature of such a procedure that the relations, patterns, formulae, geometrical figures...arrived at in this way very often turn out quite unexpectedly to apply to material settings widely different from those from which they were originally abstracted. The mathematical pattern or formula all of a sudden brings order into a domain for which it was not intended and which was never thought of when the mathematical pattern was derived. Such experiences are very impressive and very apt to create the belief in the mystical power of mathematics. 'Mathematics' appears to be at the bottom of everything, since we find it unexpectedly where we have not put it in. The fact must have struck young adepts again and again; it returns as a momentous event in the progress of physical science, as when —to give at least one famous example—Hamilton discovered that the motion of a general mechanical system was governed by exactly the same laws as a ray of light propagated in an inhomogeneous medium. Science has now become sophisticated, it has learned to be cautious in such cases, and not to take for granted an intrinsic cognateness where there may be only a formal analogy, resulting from the very nature of mathematical thought. But in the infancy of the sciences, rash conclusions, of the mystical nature characterized above, must not astonish us.

An amusing, if irrelevant, modern case of a pattern applying to an entirely different setting is the so-called transition curve in road-planning. The bend that connects two straight parts of the road ought not to be simply a circle. For this would mean that a motorist has to jerk the steering wheel suddenly as he enters the circle from the straight. The condition for an ideal transition curve presents itself: it ought to require a uniform rate of turning of the steering wheel in the first half, and the same uniform rate of turning it back in the second half of the transition. The mathematical formulation of this condition leads you to demand that the curvature must be proportional to the length of the curve. It turns out that this is a curve of very special character which was known long before the advent of motor cars, namely Cornu's spiral. Its sole application, as far as I know, had been a simple, special problem in optics, namely the interference pattern behind a slit illuminated by a point source; this problem had led to the theoretical discovery of Cornu's spiral.

A very simple problem, known to every schoolboy, is that of intercalating between two given lengths (or numbers) p and q a third, x, such that p has to x the same ratio as x to q.

$$p:x=x:q. \tag{1}$$

The quantity x is then called the 'geometrical mean' of p and q. For instance, if q were 9 times p, x would have to be 3 times p and thus one-third of q. From this you see by an easy generalization that the square of x equals the product pq,

$$x^2=pq. \tag{2}$$

(This could also be inferred from the general rule of proportions that the product of the 'inner' members equals the product of the 'outer' members.) The Greeks would interpret this formula geometrically as the 'quadrature of the rectangle', x being the side of the square whose area is the same as that of the rectangle with sides p and q. They knew algebraic formulae and equations only in geometric interpretation, since as a rule there was no *number* to fit in with the formula. For instance if you take q to be $2p$, $3p$, $5p$, ... (and p, for simplicity just 1), then x is what we call $\sqrt{2}$, $\sqrt{3}$,

$\sqrt{5}$, ..., but to them these were not numbers, they had not invented them yet. Any geometrical construction realizing the above formula is thus a geometrical extraction of the square root.

The simplest way is to plot p and q along a straight line, then erect a perpendicular at the point where they join, (N), and cut it (at C) by a circle, drawn from the centre O (the middle point of $p+q$) passing through the end points A and B of $p+q$

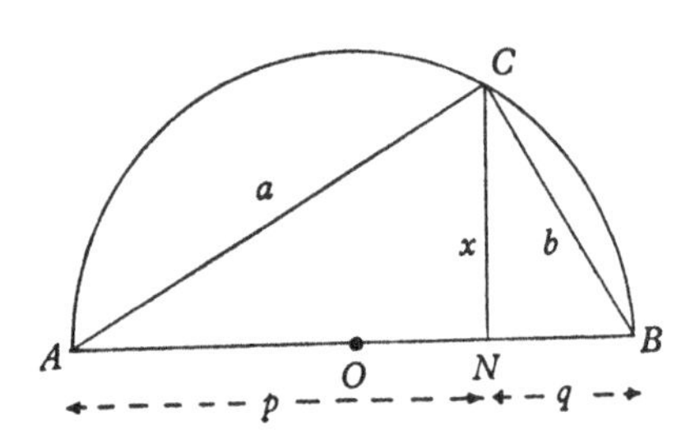

Fig. 1.

The proportion (1) then follows from the fact that ABC is a rectangular triangle, C being the 'angle in the semi-circle'; which makes the *three* triangles ABC, ACN, CNB geometrically similar. Two more 'geometrical means' are exhibited in our triangles, namely—putting $p+q=c$, the hypotenuse—

$$q : b = b : c, \qquad \text{thus } b^2 = qc,$$
$$p : a = a : c, \qquad \text{thus } a^2 = pc.$$

It follows that

$$a^2 + b^2 = (p+q)c = c^2,$$

which is the simplest proof of the so-called Pythagorean theorem.

The proportion (1) might well have occurred to the Pythagoreans in an entirely different setting. If p, q, x are lengths which you delimit on the same string by supports, or just by pressure of the finger as the violinist does, then x produces a tone 'in the middle' of those produced with p and q; the musical intervals from p to x and from x to q are the same. This may easily lead one to the problem of dividing a given musical interval into more than two equal steps. At first sight this seems to lead away from harmony, since even if the original ratio $p : q$ was rational, the intercalated steps would not be. Yet precisely this way of intercalation is followed in the equal-tempered piano-tuning, with twelve steps. It is a compromise, condemnable from the point of view of pure harmony, but hardly avoidable in an instrument with prefabricated tones.

Archytas (known also by his friendship with Plato in Tarentum around the middle of the fourth century) solved geometrically the next case, of finding *two* geometrical means (δύο μέσας ἀνὰ λόγον εὑρεῖν), or dividing a musical interval into three equal steps. This, on the other hand, also amounts to finding geometrically the third root of the given ratio q/p. In the latter form—extracting a third root—it was known as the Delian problem; Apollo's priests on the isle of Delos had once charged an oracle petitioner to double the size of their altar stone. Now this stone was a cube, and a cube of double the volume would have to have an edge $\sqrt[3]{2}$ times the given one.

In modern symbols the problem reads

$$p : x = x : y = y : q, \qquad (3)$$

from which you deduce in the above way

$$x^2 = py, \quad xy = pq. \qquad (4)$$

Multiplying member by member and cancelling the factor y:

$$x^3 = p^2 q = p^3 \frac{q}{p} \qquad (5)$$

$$x = p \sqrt[3]{\frac{q}{p}}.$$

Archytas' solution amounts to repeating the construction indicated above,

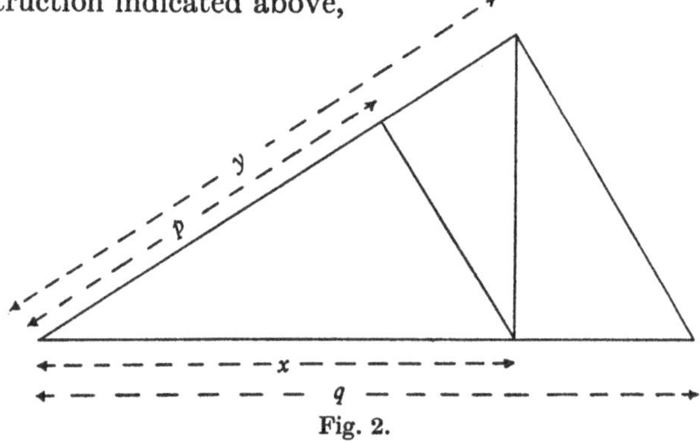

Fig. 2.

but using the second type of proportion mentioned above, which here amounts to

$$p : x = x : y \quad \text{and} \quad x : y = y : q.$$

However, this is only the final outcome of Archytas' construction, which is a very elaborate one in space, using intersections of a sphere, a cone and a cylinder—so complicated indeed that in my (first) edition of

Diels's *Presocratics* the figure that purported to illustrate the text was entirely wrong. Indeed, the above apparently simple figure cannot be constructed directly with compass and ruler from the given data, p and q. The reason is that with a ruler you can only draw straight lines (curves of the first order), with a compass only a circle, which is a particular curve of the second order; but, to extract a *third* root, a given curve of at least the *third* order has to be available. Archytas supplies it most ingeniously by those curves of intersection. His method of solution is not, as one might believe, an over-complication, but a great feat, which he achieved about half a century *before* Euclid.

The last point in Pythagorean teaching that we are going to consider here is their cosmology. It is of particular interest to us because it discloses the unexpected efficiency of an outlook, so encumbered with unfounded, preconceived ideals of perfection, beauty and simplicity.

The Pythagoreans knew that the earth was a sphere, and they were probably the first to know that. The inference was most likely drawn from its circular shadow on the moon at lunar eclipses, which they interpreted more or less correctly (see below). Their model of the planetary system and the stars is schematically and summarily indicated by the following figure.

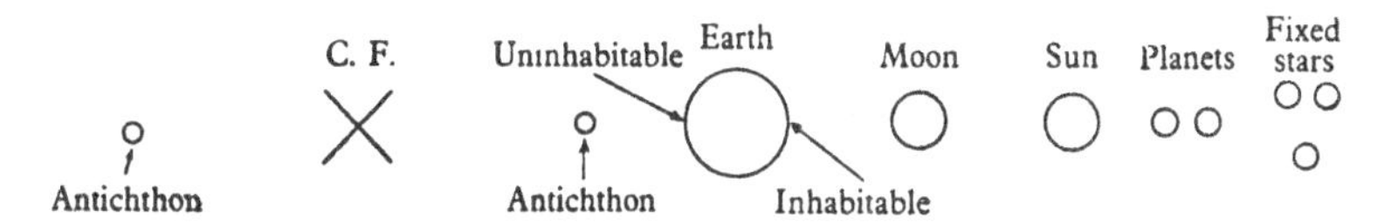

Fig. 3.

The spherical earth revolves in twenty-four hours around a fixed centre, C.F. (the *central fire, not* the sun !), to which centre it turns always the same hemisphere—as the moon does to us—which is *not habitable*, because it is too hot. Nine spheres, all centred on C.F., are imagined carrying, as it were, (1) the earth, (2) the moon, (3) the sun, (4–8) the planets, (9) the fixed stars, around the centre, each revolving at a rate peculiar to it. (Thus the lining-up along a straight line as in our figure is purely schematic; it would never arise.) There is still a tenth sphere, or at least a tenth body, the antichthon or counter-earth, of which it is not quite clear whether it was in permanent conjunction or opposition to the earth with respect to the central fire. (Our figure pictures both alternatives.) At any rate those three—earth, central fire, counter-earth—*were* supposed to be always on a straight line—naturally, since the antichthon was never seen; it was a gratuitous invention. It may have been invented for the sake of the holy number ten, but was also made responsible for such eclipses of the moon as came to pass when both sun and moon were visible at opposite points very near the horizon. This is possible because, on account of the refraction of the rays in the atmosphere, we see a star setting when it has actually been already for a few minutes below the horizon. Since this was not known, such eclipses may have presented a difficulty, which contributed to the need both for the invention of the antichthon and for the assumption that not only the moon but also the sun, the planets and the fixed stars were illuminated by the central fire and that eclipses of the moon were produced by the shadow of the earth or the antichthon in the light of the central fire.

At first sight this model appears to be so wrong that it seems hardly worth devoting any thought to it. But let us consider it carefully and remember that nothing was known about the dimensions of (*a*) the earth and (*b*) the orbits. The then known part of the earth, the Mediterranean region, actually does swing on a circle in twenty-four hours round an invisible centre, to which it always turns the same side. This, precisely, causes the rapid diurnal motion shared by all the celestial bodies. To recognize it as a merely *apparent* motion is in itself a great achievement. The point that was wrong about the motion of the earth—that in addition to the rotation it was allotted a revolution *of the same period*—was wrong only as regards the period and the centre of the revolution. These mistakes, crude as they appear to us, weigh little against the spectacular recognition that the earth is allotted the role of one of the planets, just as the sun and moon and the five that we call planets. This is an admirable feat of self-liberation from the prejudice that man and his abode needs must be in the centre of the universe—the first step towards our present outlook, which reduces our globe to one of the planets of one of the stars in one of the galaxies of the universe. It is known that this step, after having been completed by Aristarchus of Samos in about 280 B.C., was then soon taken back and the prejudice restored, to last—at least officially in some quarters—till early in the nineteenth century.

One may ask why this central fire was invented at all. The trouble of explaining those exceptional eclipses, with both sun and moon visible, would hardly have sufficed.[1] That the moon has no light of its own, but is illuminated from another source, is very early knowledge. Now, the two most impressive phenomena in the skies, the sun and the moon, are very much alike in their diurnal motions, in shape and in size; the latter is due to the coincidence that the moon is about as many times nearer to us as it is smaller. This necessarily induces one to put the two on the same footing, to transfer what is known about the moon to the sun, and thus to consider them both illuminated from the same source, which is just the hypothetical central fire. But since it was not seen, there was no other place to put it but 'under our feet', covered to our eyes by our own planet.

This model, though perhaps wrongly, is ascribed to Philolaus (second half of the fifth century). A glance at its further development goes to show that even gross mistakes, made under the bias of preconceived ideas about perfection and simplicity, can be relatively innocuous; nay, the more arbitrary and unfounded such an assumption is, the less mental damage will it do, for experience will eliminate it the more rapidly. As has once been said, a wrong theory is better than none at all.

In the present case, first the travels of Carthaginian merchants, extending beyond the 'pillars of Hercules', and a little later Alexander's expedition to India disclosed nothing about the central fire or the antich-

[1] It is, by the way, not sure that such an eclipse ever has been observed.

thon, or about the earth getting less habitable beyond the limits of the Mediterranean culture. So all this had to be dropped. With the fictitious centre (the central fire) gone, it was natural to abandon the idea of the earth's diurnal revolution and to replace it by a pure rotation around its own axis. There is dissension among the historians of ancient philosophy in deciding to whom the 'new doctrine of the rotation of the earth' is due; some say to Ecphantus, one of the youngest Pythagoreans, others are inclined to regard him only as a personage in a dialogue of Heraclides Ponticus (a native of Heraclea on the Black Sea, who attended the schools of Plato and Aristotle) and to ascribe this 'new doctrine' (which, by the way, Aristotle mentions but rejects) to Heraclides. But it is perhaps more relevant to emphasize that there is no question of a new doctrine; the rotation of the earth was already contained in Philolaus' system: a body that revolves around a centre and keeps turning always the same side towards it—as the moon does with respect to the earth—must not be said to have no rotation, but to rotate with a period exactly equal to its period of revolution. This is not a sophisticated scientific description, nor is the equality of periods in the case of the moon (and others similar to it) a haphazard coincidence; it is due to tidal friction either in a previously existing oceanic or atmospheric cover on the moon or in the bulk of its body.[1]

[1] The tidal friction on the *earth* produces a (*very* slow) retardation of its rotation. The reaction on the moon is bound to be a (*very* slow) recession from the earth together with a corresponding increase in the moon's period of revolution. From this, one is inclined to conclude that there must be even now some weak agent at work to maintain the exact equality of the two periods of the moon.

Now as we stated above, Philolaus' system did ascribe to the earth, with respect to the central fire, exactly this kind of motion, a rotation and a revolution with the same period. To drop the latter does not amount to the discovery of the former, since it was already discovered. We are rather inclined to call it a step in the wrong direction, for revolution there is, albeit around another centre.

But the aforesaid Heraclides, who was in intimate contact with the later Pythagoreans, has to be credited, so it seems, with the most momentous step towards recognizing the actual situation. The striking changes of brightness of the inner planets, Mercury and Venus, had been noticed. Heraclides correctly attributed them to their changing distance from the *earth*. Hence they could not move in circles around the latter. The further fact that in their main or average motion they follow the sun's course probably helped to prompt the correct view that those two anyhow move in circles around the sun. Similar considerations would soon apply to Mars which also exhibits considerable changes in brightness. Eventually, as is well known, Aristarchus of Samos established (about 280 B.C.) the heliocentric system, only about one century and a half after Philolaus. Its soundness was not recognized by many, and about another 150 years later it was overthrown by the authority of the great Hipparchus 'President of the University of Alexandria', as he would be called in our days.

It is an amazing fact, not a little disconcerting to the sober scientist of today, that the Pythagoreans with all their prejudices and preconceived ideas about beauty and simplicity made better headway, at any rate in this one important direction, towards an understanding of the structure of the universe—better than the sober school of Ionian 'physiologoi', of whom we shall have to speak presently, and better than the atomists who succeeded them spiritually. For reasons that will appear soon scientists are very much inclined to regard the Ionians (Thales, Anaximander, etc.) and, above all, the great atomist Democritus as their spiritual ancestors. Yet even the last-named clung to the idea of a flat, tambourine-shaped earth, which was perpetuated among the atomists by Epicurus and lasted down to the poet Lucretius, in the first century B.C. Disgust at the unfounded, weird phantasies and the arrogant mysticism of the Pythagoreans may have contributed to cause a clear thinker like Democritus to reject all their teaching that gave the impression of arbitrary, artificial construction. Yet their power of observation, trained in those early, simple acoustical experiments about vibrating strings, must have enabled them to recognize through the fog of their prejudices, something so near the truth that it served as a good foundation from which the heliocentric view rapidly sprang. Sad to say, it was equally rapidly discarded under the influence of the Alexandrian school, who believed themselves to be sober scientists, free of prejudice, guided only by facts.

I have not mentioned in this brief survey the anatomical and physiological discoveries of Alcmaeon of Croton. He was a younger contemporary of Pythagoras;

he discovered the main sensual nerves and followed their course to the brain, which he recognized as the central organ corresponding to the activity of the mind. Up to then—and, in spite of his discovery, for a long time after—the heart (ἦτορ, καρδία), the diaphragm (φρένες) and breathing (πνεῦμα, Lat. *anima* > *animus*) were deemed to be connected with the mind or the soul, as is evidenced by the expressions that were used metaphorically to indicate them. Vestiges of these metaphors are to be found in all modern languages. But let this suffice for our present purpose. The reader can easily find elsewhere more competent information on the medical achievements of antiquity.

CHAPTER IV

THE IONIAN ENLIGHTENMENT

Turning now to the philosophers usually classed together under the name of the Milesian School (Thales, Anaximander, Anaximenes) and, in the next chapter, to some more or less connected with them (Heraclitus, Xenophanes), then to the atomists (Leucippus, Democritus), let me point out two things. First, the order in regard to the preceding chapter is not chronological; the *floruit* of the three Ionian 'physiologoi' (Thales, Anaximander, Anaximenes) is approximately dated at 585, 565, 545 B.C. respectively, as against Pythagoras 532 B.C. Secondly, I wish to point out the double role that this whole group plays in our present context. They are a group of definitely scientific outlook and aims, just as the Pythagoreans were, but opposed to them as regards the competition 'Reason *v.* Senses', explained in our second chapter. They take the world as given to us by our senses and try to explain it, not bothering about the precepts of reason any more than the man in the street does, from whose way of thinking theirs is a direct descendant. Indeed it frequently starts from problems or analogies of handicraft and serves practical applications in navigation, mapping, triangulation. On the other hand let me remind the reader about *our* main problem, which will be to find out the special and somewhat artificial features of present-day science that are supposed (Gomperz,

Burnet) to originate from Greek philosophy. We shall submit and discuss two such features, namely the assumption that the world *can be understood,* and the simplifying provisional device of *excluding the person of the 'understander'* (the subject of cognizance) from the rational world-picture that is to be constructed. The *first* definitely originated from the three Ionian 'physiologoi', or from Thales, if you like. The *second,* the exclusion of the subject, has become an ingrained habit of old. It became inherent in any attempt to form a picture of the objective world such as the Ionians made. So little was one aware of the fact that this exclusion was a special device, that one tried to trace the subject *within* the material world-picture in the form of a soul, whether a material one, made of particularly fine, volatile and mobile matter, or a ghostlike substance that interacts with matter. These naïve constructions went down through the centuries and are far from extinct today. Though we cannot trace the 'exclusion' as a definite step, decided upon consciously (which it probably never was), we do find in the fragments of Heraclitus (*floruit* about 500 B.C.) remarkable evidence of his being aware of it. And the fragment of Democritus which we have already quoted at the end of Chapter II shows him worried about the fact that his atomistic model of the world is devoid of all the subjective qualities, the sensorial data, from which it was built.

The movement called the Ionian enlightenment began in that very remarkable sixth century B.C.; indeed, it so happened that in this century also in the

Far East spiritual trends of tremendous consequence were started, connected with the names of Gautama Buddha (born about 560 B.C.), Laotse and his younger contemporary Confucius (born 551 B.C.). The Ionian group sprang up, ostensibly out of nothing, in the narrow fringe called Ionia, the west coast of Asia Minor and the islands in front of it. The particularly favourable geographical and historical conditions obtaining there and then have often been depicted in more splendid rhetoric than I have at my disposal; the situation was favourable to the development of free, sober, intelligent thought. Let me mention three points.

The region (like southern Italy in the time of Pythagoras) did not belong to a big powerful state or empire, such as is usually found hostile to free thinking. It consisted politically of many small, self-governed and well-to-do city- or island-states, either republics or tyrannies. In either case they seem to have been ruled or governed quite frequently by *the best brains*, which has at all times been a rather exceptional event.

Secondly, the Ionians, inhabiting islands and the very broken coast of the mainland, were a seafaring people, interposed between East and West. Their flourishing trade mediated the exchange of goods between the coasts of Asia Minor, Phoenicia and Egypt on the one side and Greece, southern Italy and southern France on the other. Mercantile exchange has always and everywhere been, and still is, the principal vehicle for an exchange of ideas. Since the persons between whom this exchange first takes place are not closet-scholars, poets or teachers of philosophy, but sailors and merchants, it is bound to take its start from practical problems. Manufacturing devices, new techniques in handicraft, means of transport, aids to navigation, methods of laying out harbours, of erecting piers and warehouses, harnessing water supply, and so on, will be among the first things one people learns from the other. The rapid development of technical skill, which results in an intelligent people from a vital process of this kind, stirs the minds of theorizing thinkers, who will often be called upon for help in carrying out some newly learned art. If they apply themselves to abstract problems about the physical constitution of the world, their whole way of thinking will show traces of the practical origin from which it started. This is precisely what we find in the Ionian philosophers.

As a third favourable circumstance it has been pointed out that these communities, to put it briefly, were not priest-ridden. There was not, as in Babylonia and Egypt, a hereditary privileged priestly caste of the kind that, if they are not themselves the rulers, usually side with them in opposing the development of new ideas, since they have an instinctive feeling that any change in outlook might eventually turn against themselves and their privileges. So much for the conditions that favoured the rise of a new era of independent thought in Ionia.

Many a schoolboy or young student may have come across, in his text-book or elsewhere, a brief survey of Thales, Anaximander, etc. On reading how one taught that everything was water, another that everything was air, a third that everything was fire, and on learning about such queer ideas as fiery channels with windows in them (the celestial bodies), the streams up and down the atmosphere, etc., he may well have been bored and have wondered why he was asked to be interested in such naïve old stuff which we know is completely beside the point. What is then the great thing that happened at that time in the history of ideas, what makes us call this event the Birth of Science and speak of Thales of Miletus as the first scientist in the world (Burnet)?

The grand idea that informed these men was that the world around them was something *that could be understood*, if one only took the trouble to observe it properly; that it was not the playground of gods and ghosts and spirits who acted on the spur of the moment and more or less arbitrarily, who were moved by passions, by wrath and love and desire for revenge, who vented their hatred, and could be propitiated by pious offerings. These men had freed themselves of superstition, they would have none of all this. They saw the world as a rather complicated mechanism, acting according to eternal innate laws, which they were curious to find out. This is, of course, the fundamental attitude of science up to this day. To us it has become flesh of our flesh, so much so that we have forgotten that somebody had to find it out, make it a programme, and embark on it. Curiosity is the stimulus. The first requirement of a scientist is to be curious. He must be capable of being astonished and eager to find out. Plato, Aristotle and Epicurus emphasize the import of being astonished (θαυμάζειν). And this is not trivial when it refers to general questions about the world as a whole; for, indeed, it is given us only once, we have no other one to compare it with.

We call this the *first step*, which was of paramount importance, quite irrespective of the adequacy of the explanations actually offered. I believe it is correct to say that it was a complete novelty. The Babylonians and Egyptians knew, of course, a lot about the regularities in the orbits of the heavenly bodies, particularly about eclipses. But they regarded them as religious secrets and were far from seeking natural explanations. And they were certainly very far from contemplating an exhaustive description of the world in terms of such regularities. In Homer's poems the incessant interference of the gods with natural events, the repelling human sacrifices reported in the *Iliad* illustrate what was said above in general terms. But to recognize the Ionians' outstanding discovery in creating for the first time a truly scientific outlook, we need not contrast them with those who had preceded them. So little did the Ionians succeed in uprooting superstition, that in all the time to come, down to our own days, there is no epoch that was not riddled with superstition. In this I am not referring to popular belief, but to the wavering attitude even of truly great men, such as Arthur Schopenhauer, Sir Oliver Lodge, Rainer Maria Rilke, to name just a few. The Ionians' attitude lived on with the atomists (Leucippus, Democritus, Epicurus, Lucretius) and with the scientific schools of Alexandria though in different ways; for, unhappily, natural philosophy and scientific research had separated in the last three centuries B.C., much as in modern times. After this the scientific outlook gradually died away, when in the first centuries of our era the world became increasingly interested in ethics and in strange brands of metaphysics, and did not care for science. Not before the seventeenth century did the scientific outlook regain momentum.

The *second step*, of almost equal moment, can also be traced back to Thales. It is the recognition that all matter of which the world consists has, with all its infinite variety, yet so much in common, that it must be intrinsically the same stuff. We may well call this Proust's hypothesis in the embryonic stage. It was the first move towards an understanding of the world, thus towards implementing what we have called the *first* step, the conviction that it can be understood. From our present outlook we must say that this move touched the most essential point and was amazingly adequate. Thales ventured to regard water (ὕδωρ) as the basic stuff. But we had better not associate this naïvely with our 'H_2O', rather with liquid or fluid (τὰ ὑγρά) in general. He may have observed that all life appeared to originate in the liquid or in the moist. In deeming the most familiar liquid (water) the *one* material of which everything was composed, he implicitly asserted that the physical state of aggregation (solid, fluid, gaseous) was a secondary affair, not very essential. We cannot expect him to be satisfied—as would befit a modern mind—by just saying: we shall just give it a name, call it *matter* (ὕλη), and investigate its properties. A new discovery is usually overstated and very often formulated into a hypothesis with too much detail that wears off later. This comes from our intense desire of 'finding out', from the urge of scientific curiosity, which is so essential for finding out anything, as we said above. A rather interesting detail, reported by several doxographers as Thales' opinion, is that the land floats on the water 'like a piece of wood'; which must mean with a considerable part of it immersed. This recalls on the one hand the old myth about the isle of Delos wandering around erratically until Leto there gave birth to twins, Apollo and Artemis; but it is also amazingly akin to the modern theory of isostasy, according to which the continents do float on a liquid, though not exactly on the water of the oceans, but on a heavier molten substance below them.

In point of fact, Thales' 'overstatement' or 'rashness' in forming his general hypothesis was soon corrected by his disciple and associate (ἑταῖρος) Anaximander, who was younger by roughly twenty years. He denied that the universal world material was identical with any known stuff and invented a name for it, calling it the Boundless (ἄπειρον). Much ado there was in antiquity about this interesting term, as if it were anything but a newly invented name. I shall not dwell upon it, but shall follow the trend of essential physical ideas by indicating what I would call *the third momentous step* in their development. It is due to Anaximenes, the associate and disciple of Anaximander, roughly another twenty years younger (died about 526 B.C.). He recognized that the most obvious transformations of matter were 'rarefaction and condensation'. He explicitly maintained that every kind of matter could be transformed into the solid, fluid or gaseous state in suitable circumstances. As the basic substance he chose to regard the air, thus treading again firmer ground than his master. Indeed, had he said 'dissociated hydrogen gas' (which he could hardly be expected to say), he would not be far from our present view. Anyhow from the air, he said, lighter bodies (viz. the fire and the lighter, purer element on the top of the atmosphere) were formed by further rarefaction, while mist, clouds, water, and the solid earth resulted from successive steps of condensation. These assertions are as adequate and correct as they could at all be formulated with the knowledge and within the conceptions of the time. Take note that it is not a question of only small changes of volume. On the transition from the ordinary gaseous state to the solid or liquid state the density increases by a factor of between one and two thousand. For instance a cubic inch of water vapour at atmospheric pressure, when condensed, shrinks to a drop of water of little more than a tenth of an inch in diameter. Anaximenes' view, that liquid water and even a firm, solid stone are formed by the condensation of a basic gaseous substance (though it seems to amount to the same as the opposite view of Thales), is yet both bolder and much more akin to our present view. For we do consider a gas to be in the simplest, most primitive, 'non-aggregated' state, from which the relatively complicated formation of liquids and solids results by the intervening of agents that play a subordinate part in the gas. That Anaximenes did not indulge in abstract phantasies, but was eager to apply his theory to concrete facts, can be seen from the amazingly correct insight he attained in some cases. Thus he tells us, concerning the difference between hail and snow (both consisting of water in the solid state, i.e. of ice), that hail is formed when the water falling down from the clouds (i.e. raindrops) freezes, while snow results from moist clouds themselves taking to the solid state. A modern text-book of meteorology will tell you nearly the same. The stars (let me mention this by the way and out of order) do not give us heat, he said, because they are so far away.

But by far the most important point about the rarefaction-condensation theory is that it was the stepping-stone to atomism which actually followed very soon in its wake. This point deserves attention, for to us modern people it is not obvious, we are too sophisticated. We are familiar with the idea of the *continuum*, or we believe ourselves to be. We are *not* familiar with the enormous difficulty this concept presents to the mind, unless we have studied very modern mathematics (Dirichlet, Dedekind, Cantor). The Greeks hit on these difficulties, became fully aware of them, were profoundly shaken by them. This can be seen from their embarrassment because 'no number' corresponds to the diagonal of a square with side 1 (we say, it is $\sqrt{2}$); it can be seen from Zeno's (the Eleatic's) well-known paradoxes about Achilles and the tortoise, and about the arrow in flight, as well as from some other paradoxes about sand, and from the recurring questions about the line consisting of points—and, if so, of how many? That we (those of us who are not mathematicians) have learned to shirk these difficulties, and have unlearned how to understand the Greek mind on this point, is, I believe, largely due to the decimal notation. At some time in our schooldays we are made to swallow the lump that one may contemplate decimal fractions whose figures run to infinity, and that such a one represents a number even when no simple recurrence of the figures can be indicated. The lump is lubricated by our having learnt a little earlier that quite simple numbers, as e.g. $\frac{1}{7}$ (one-seventh), have no finite decimal fractions corresponding to them, but infinite ones, *with recurrence*:

$$\tfrac{1}{7} = 0{\cdot}142857 \mid 142857 \mid 142857 \mid \ldots.$$

The enormous difference between this case and, say,

$$\sqrt{2} = 1{\cdot}4142135624 \ldots,$$

appears when we reflect that the $\sqrt{2}$ would conserve its character whatever 'basis' we chose[1] instead of our conventional basis 10, whereas with the basis 7 we, of course, get for $\frac{1}{7}$ the 'septimal fraction'

$$\tfrac{1}{7} = 0{\cdot}1.$$

Anyhow, after we have swallowed the lump, we feel that we are now in a position to ascribe a definite number to any point on the straight line between zero

[1] The square root of 2 in septimal fractions reads: $1{\cdot}2620346 \ldots.$

and one, or indeed between zero and infinity, or indeed between minus infinity and plus infinity, if a point zero is marked on it. We feel in possession and in control of the *continuum.*

In addition we know india-rubber. We know that we can stretch a string of india-rubber within large limits, or even a rubber surface, when blowing up a child's balloon. We have no difficulty in imagining that we can do a similar thing with a solid rubber body. And so we have no difficulty in reconciling a continuous model of matter with even very considerable changes of its shape and volume, though indeed quite a few physicists in the nineteenth century found some difficulty in doing so.

The Greeks, for the reasons just mentioned, had not this facility. They were bound to interpret the volume changes sooner or later in the way that the bodies consist of discrete particles, which themselves do not change, but recede from each other or come closer together, leaving more or less empty space between them. That was their, and that is our, atomic theory. It might seem that a deficiency—the lack of knowledge about the continuum—just happened to lead them the right way. Fifty years ago one could still have accepted this conclusion, in spite of its intrinsic improbability. The latest phase of modern physics, inaugurated in 1900 by Planck's discovery of the quantum of action, points in the opposite direction. While accepting the atomism of ordinary matter from the Greeks, we seem still to have made an improper use of our familiarity with the continuum. We have used this concept for *energy*: but Planck's work has cast doubt on its adequacy. We still use it for space and time; it will hardly ever be dropped in abstract geometry; but it may very well turn out to be out of place for physical space and physical time. So much for the development of physical ideas by the Milesian school, which is, I believe, their most important contribution to Western thought.

A well-known statement about them is that they deemed all matter to be alive. Aristotle, dealing with the soul, tells us that some people considered it mixed up with 'the whole', and that in this way Thales thought everything to be full of gods; that he attributed some moving power to the soul and ascribed a soul even to the stone, because it moved the iron. (This refers, of course, to the loadstone.) This and the similar property imparted to amber (ḗlektron) when electrified by rubbing are given elsewhere as the reasons why Thales ascribed a soul even to the inanimate (=soulless). Again, it is said that he regarded God as the intellect (or mind) of the universe and thought the whole to be animate (endowed with soul) and full of deities. The name of 'hylozoists' (hýlē, matter; zō-ós, alive) for the Milesian school was invented in later antiquity to indicate their view on this point, which must have seemed rather odd and childish then. For Plato and Aristotle had stipulated a clear distinction between the alive and the inanimate: the alive is what moves itself, e.g. man, a cat or a bird, the sun, moon and the planets. Some modern views approach closely to what the hylozoists meant and felt. Schopenhauer extended his fundamental notion of 'Will' to everything, he ascribed will to the falling stone and to the growing plant as well as to the spontaneous motions of animals and man. (He regarded conscious cognizance and intellect as a secondary, accessory phenomenon, a view with which this is not the place to quarrel.) The great psycho-physiologist G. Th. Fechner entertained, albeit in his hours of leisure, ideas about the 'souls' of the plants, the planets, the planetary system, that are interesting to read and were meant to convey a little more than just diverting day-dreams. Finally, let me quote from Sir Charles Sherrington's Gifford Lectures 1937–8, published in 1940 as *Man on his Nature.* A discussion of many pages on the physical (energetical) aspect of material events, and of the doings of organisms in particular, is summarized by pointing out the historical position of our present outlook thus: '...in the Middle Ages, and after them..., as with Aristotle before, there was the difficulty of the animate and the inanimate and the finding of the boundary between them. Today's scheme makes plain why that difficulty was, and dissolves it. There is no boundary.'[1] If Thales could read this, he would say: 'This is just what I held two hundred years before Aristotle.'

The idea that organic and inorganic nature form an inseparable union did not remain a barren philosophical statement with the Milesians, as it did, for example, with Schopenhauer, whose chief mistake was that he opposed (or, perhaps better, he ignored) *evolution,* though biological evolution was, in Lamarck's version, established at his time and had a great influence on

[1] 1st ed., p. 302.

some contemporary philosophers. In the Milesian school one immediately drew the consequences, taking it for granted that life must originate from lifeless matter somehow, and obviously in a gradual way. We mentioned above that Thales decided on water as the primordial substance probably because he thought he witnessed life originating spontaneously in the wet or moist. In this he was, of course, mistaken. But his disciple Anaximander, pondering on the origin and development of living beings, arrived at remarkably correct conclusions, and, what is more, by remarkably sound observation and inference. From the helplessness of newborn land-animals, including human babies, he concluded that this cannot have been the earliest form of life. Fishes, however, usually give no further attention to the progeny that proceeds from their spawn. Their young ones have to get on alone, and—we may add—they can manage more easily, because gravity is compensated in the water. Life must therefore have come out of the water. Our own ancestors were fishes. All this coincides so remarkably with modern findings, and is so intrinsically sound, that one regrets the romantic details that are added. Certain fishes, perhaps a kind of shark (γαλεός), were believed—in contrast to what we said just before—to nurse their young ones with particular tenderness, indeed to keep them in (or even to take them back into) their wombs until they had reached a stage where they were fully capable of supporting themselves. Anaximander is said to have maintained that such a type of child-loving fish were our ancestors, in whose wombs we developed until we were able to get out on the land and survive there for a certain time. Reading this romantic and illogical story one cannot help remembering that most of our reports, if not all, are by writers who heartily disagreed with Anaximander's theory, which had been ridiculed rather unfairly by the great Plato. They were therefore hardly disposed to understand it. Might it be that Anaximander pointed out, very consistently, an intermediate stage between fish and land animals—namely the Amphibia (the class to which the frogs belong) which spawn in the water, begin their lives in the water, then after a considerable metamorphosis come out on the land to live there for a while? Somebody who found it all too ridiculous, that a fish should gradually develop into a man, could easily distort this into that 'explanatory' story, which makes man grow *inside* a fish. It bears quite a family likeness to other romantic fictions on natural history with which the Socrates-Plato circle used to amuse themselves.

CHAPTER V

THE RELIGION OF XENOPHANES. HERACLITUS OF EPHESUS

The two great men of whom I wish to tell in this section have this in common, that they both give you the impression of walkers-alone—deep original thinkers, influenced by others, but not pledged to any 'school'. The most probable period for Xenophanes' life is the century after about 565 B.C. At the age of ninety-two he describes himself as having wandered through the Greek countries (including, of course, Magna Graecia) for the last sixty-seven years. He was a poet, and the fragments of his fine verse that have come down to us make one deeply regret that his, as well as Empedocles' and Parmenides', hexameters and elegiacs were mostly lost, while the war-songs of the *Iliad* were preserved. Even so, what is extant of all these philosophical poems would in my opinion make a more interesting, a worthier and a more suitable subject for our school reading than the Wrath of Achilles (if you think what it is about).[1] According to Wilamowitz, Xenophanes 'upheld the only real monotheism that has ever existed upon earth'.

He was the same who discovered and correctly *interpreted fossils in the rocks of south Italy—in the* sixth century B.C.! I wish to quote here some of his

[1] I do not wish it to be inferred that I regard the *Iliad* as nothing but a war-song whose loss would not have been deeply deplorable.

famous fragments that give us an idea of the attitude of the advanced thinkers of that period towards religion and superstition. To make room for a scientific view of the world, it was, of course, necessary first to clear away the ideas of Jove raising thunder and throwing thunderbolts, of Apollo causing a pestilence to vent his anger, etc.

Xenophanes remarks (fr. 11)[1] that Homer and Hesiod ascribe to the gods all things that are a shame and disgrace among mortals, imposture, stealing and adultery and deceiving one another with great ingenuity. And (fr. 14): 'Mortals deem that the gods are begotten as they are and have clothes like theirs and a voice and form.'

Let me stop for a moment to ask: How could the general Greek public accept such a low idea of the gods? The answer is, I think, that to them it did not seem low at all. On the contrary, it testified to the gods' power and freedom and independence that they were allowed to do things blamelessly which we are blamed for, because we are poor little mortals only. They shaped their gods in the image of the great and rich and mighty and powerful and influential people among them, who most likely then as now could afford to evade the law and indulge in crime and shameful deeds, on the strength of their power and wealth.

In several fragments Xenophanes dethrones the gods in a couple of lines by ridiculing them as being patently nothing but the product of human imagination.

[1] The numbering of the fragments follows Diels' first edition throughout.

(Fr. 15) Yes, and if the oxen or horses or lions had hands and could paint with their hands, and produce works of art as men do, horses would paint the forms of the gods like horses, and oxen like oxen and make their bodies in the image of their several kinds.

(Fr. 16) The Ethiopians make their gods black and snub-nosed; the Thracians say theirs have blue eyes and red hair.

Then a few short fragments, giving us his own idea about the godhead—clearly in the singular:

(Fr. 23) One god, the greatest among gods and men, neither in form like unto mortals nor in thought.

(Fr. 24) He sees all over, thinks all over, and hears all over.

(Fr. 25) But without toil he swayeth all things by the thought of his mind.

(Fr. 26) And he abideth ever in the selfsame place, moving not at all; nor does it befit him to go about now hither, now thither.

And then his, to me, particularly impressive agnosticism:

(Fr. 34) There never was nor will be a man who has certain knowledge about the gods and about all the things I speak of. Even if he should chance to say the complete truth, yet he himself knows not that it is so. It is all nought but chancing opinion.

Let us turn to a slightly later thinker, Heraclitus of Ephesus. He was a little younger (flourished around 500 B.C.); probably *not* a disciple of Xenophanes, but acquainted with his writings and influenced by him and by the older Ionians. He already passed for 'obscure' in antiquity and was, I daresay, for this reason seized upon by Zeno, the founder of the Stoic school, and by the later Stoics, including Seneca. The few extant fragments bear witness to this. The details of his physical world-picture are of little interest. The general trend of ideas was that of Ionian enlightenment, with a strong agnostic tinge, akin to Xenophanes. Some plain and characteristic statements are:

(Fr. 30) This world, the same for all of us, none of the gods nor of the humans has made; it has always been and is and will be, an ever living fire, flaring up in parts, in parts dying down.

(Fr. 27) There awaits men when they die such things as they look not for nor dream of.

As an example of the obscure fragments (the translation is Burnet's):

(Fr. 26) Man kindles a light for himself in the night-time, when he has died but is alive. The sleeper, whose vision has been put out, lights up from the dead; he that is awake lights up from the sleeping.

A group of fragments seem to me to point to very deep epistemological insight, namely this: since all knowledge is based on sense perceptions, these must be, *a priori*, equally valued, whether they occur in waking or dream or hallucination, whether in a person of sound mind or not so. What makes the difference and enables us to build up a reliable world-picture from them is that this world can be so constructed as to be *in common* to all of us, or rather to all *waking*, *sane* persons. (You must not forget that at the time it was much more usual to think of apparitions in dreams as something real; Greek mythology is full of that sort of thing.) Those fragments read:

(Fr. 2) It is therefore necessary to follow *the common*. But while reason (λόγος) *is* common, the majority live as though they had a private insight of their own.

(Fr. 73) We must not act and speak like sleepers. (Explanation: for then (in our sleep) too we believe that we act and speak.)

And mainly:

(Fr. 114) Those who speak with a sound mind (ξὺν νόῳ) must hold fast to what is common to all, just the same as a city holds on to her law, nay much more strongly so; for all the laws of men are fed by the one divine law. This prevails as much as it will and suffices for all things with a net surplus.

(Fr. 89) The waking have one common world, but the sleeping turn aside each into a world of his own.

What impresses me particularly is the great emphasis on holding fast to what is common—viz. to escape insanity, to escape being an 'idiot' (from ἴδιος, private, one's own). He was not a socialist—if anything an aristocrat, maybe a 'fascist'.

I believe this interpretation to be right. Nowhere could I find a reasonable explanation for this 'common' in a man like him. He says once something like this: one man of genius weighs more than ten thousand of the crowd. He reminds one sometimes quite strongly of Nietzsche—the great 'fascist'! All good things are brought about by strife and struggle.

To sum up, the meaning is, I think, that we form the ideas of a real world around us from the fact that part of our sensations and experiences overlap, as it were; this overlapping part—that is the real world.

Generally speaking one ought not, I think, to be altogether too astonished to find occasionally very deep philosophical thought in the earliest records of human thinking about the world; to find ideas which to form or to grasp costs us nowadays some effort and labour of abstraction. One may think this infancy of human thought is, figuratively, 'still nearer to Nature'. The rational picture of the world was not yet attained, the construction of 'the real world around us' not yet achieved. At any rate we do have many instances of such early deep thought in the old religious writings of many peoples, the Indians, the Jews, the Persians.

In comparing these early periods of deep philosophical awareness, I cannot help remembering a word of P. Deussen, the great Sanskritist and interesting philosopher, who said: 'It is a great pity that children in the first two years of their life cannot talk, for if they could, they probably would talk Kantian philosophy.'

CHAPTER VI

THE ATOMISTS

Is the ancient atomic theory, which is attached to the names of Leucippus and Democritus (born around 460 B.C.), the true forerunner of the modern one? This question has often been asked and very different opinions about it are on record. Gomperz, Cournot, Bertrand Russell, J. Burnet say: Yes. Benjamin Farrington says that it is, 'in a way', and that the two have a lot in common. Charles Sherrington says: No, pointing to the purely qualitative character of ancient atomism and to the fact that its basic idea, embodied in the word 'atom' (uncuttable or indivisible), has made this very name a misnomer. I am not aware that the negative verdict has ever passed the lips of a classical scholar. And when it comes from a scientist, he always shows by some remark that he regards chemistry —not physics—as the proper domain of the notions of atoms and molecules. He will mention the name of Dalton (born 1766) and omit, in this context, the name of Gassendi (born 1592). It was the latter who definitively reintroduced atomism into modern science, and he came to it after studying the fairly substantial extant writings of Epicurus (born around 341 B.C.), who had taken up the theory of Democritus, of which only scarce original fragments have come down to us. It is noteworthy that in chemistry, after the momentous development that had followed the discoveries of

Lavoisier and Dalton, a strong movement ('energeticists'), headed by Wilhelm Ostwald, supported by the views of Ernst Mach, arose towards the end of the nineteenth century in favour of abandoning atomism. It was said that it was not needed in chemistry and ought to be dropped as an unproved and unprovable hypothesis. The question as to the origin of ancient atomism and as to its connexion with modern theory is of much more than purely historical interest. We shall return to it. First, I shall briefly indicate the main features of Democritus' views. They are these:

(i) The atoms are invisibly small. They are all of the same stuff or nature (φύσις), but there is an enormous multitude of different shapes and sizes, and that is their only characteristic property. For they are impermeable and act on each other by direct contact, pushing and turning each other; and thus the most various forms of aggregation and interlacing of atoms of the same and of different kinds produce the infinite variety of material bodies, as we observe them, in their manifold interaction with each other. The space outside them is empty—a view that seems natural to us, but was subject to infinite controversy in antiquity, because many philosophers concluded that the μὴ ὄν, the thing that *is not*, could not possibly *be*, that is to say there cannot be empty space!

(ii) The atoms are in *perpetual motion*, and we may take it that this motion was regarded as irregularly or disorderly distributed in all directions, since nothing else is thinkable if the atoms are to be in perpetual motion even in bodies that are at rest or move with slow speed. Democritus states explicitly that in empty space there is no above and below, in front or behind, no direction privileged—empty space is isotropic, we should say.

(iii) Their continual motion persists by itself, it does not come to rest; this was taken for granted. This discovery, by guess, of the *law of inertia* must be regarded as a great feat, since it is patently contradicted by experience. It was reinstated 2000 years later by Galileo, who arrived at it by ingenious generalization on carefully conducted experiments about pendulums and balls rolling down an inclined groove. At the time of Democritus it did not seem at all acceptable; it gave great difficulty to Aristotle, who regarded only the circular motion of the celestial bodies as a natural one that could persist without change indefinitely. In modern terms we should say that the atoms were endowed with *inertial mass*, which made them continue their motion in empty space and impart it to other atoms which they hit.

(iv) Weight or gravity was *not* regarded as a primitive property of the atoms. It was explained in a manner that in itself is quite ingenious, namely by a general whirl-motion which makes the bigger, more massive atoms tend towards the centre where the rotational speed is smaller, the lighter ones being thereby pushed or thrown away from the centre, into the heavens. Reading the description one is reminded of what happens in a centrifuge, though this, of course, is quite the opposite, the specifically heavier bodies being thrust outward, the lighter ones tending towards the centre. On the other hand if Democritus had ever made himself a cup of tea and stirred it round with a spoon, he would have found the tea leaves gathering in the centre of the cup, an excellent example to illustrate his whirl-theory. (The true ground of this is again just the opposite, the whirl being stronger in the middle than in the outer parts where it is retarded by the walls.) What amazes me most is this: one would think that this idea of gravity being due to a continual whirl would automatically suggest a world-model of spherical symmetry, and thus a spherical earth. But that was not the case, Democritus rather inconsistently kept to the form of a tambourine; he continued to regard the daily revolutions of the celestial bodies as real—and let the tambourine-earth reside on an air-cushion. Perhaps he was so disgusted by silly talk of the Pythagoreans and Eleatics that he did not wish to accept anything from them.

(v) But, to my view, the gravest defeat the theory suffered, which condemned it to become a 'sleeping beauty' for so many centuries, was due to its extension to the *soul*; the soul was considered as composed of material atoms, particularly fine ones with particularly high mobility, probably spread all over the body and attending its functions. This was sad, because it was bound to repel the finest and deepest thinkers in the following centuries. We must be careful not to take Democritus to task too severely. It was thoughtlessness in a man whose deep understanding of the theory of knowledge I shall prove presently. He took over, and implemented along the lines of the atomic theory, the old misconception, firmly anchored in the language up to the present day, that the soul is breath. All the old words for soul originally meant air or breath: ψυχή, πνεῦμα, *spiritus*, *anima*, (Sanskr.) *athman* (modern: expire, animate, inanimate, psychology, etc.). Well, this breath was air, and air was composed of atoms—and so the soul was composed of atoms. It is a condonable short-cut to the central metaphysical problem, which really is unsolved up to the present day—see the masterful discussion in Charles Sherrington's *Man on his Nature*.

It has a terrible consequence, which has haunted the thinkers of many centuries and in slightly changed form still puzzles us today. The world-model consisting of atoms and empty space implements the basic postulate of *Nature being understandable*, provided that at any moment the subsequent motion of the atoms is uniquely determined by their present configuration and state of motion. Then the situation reached at any moment engenders of necessity the following one, and this the next following one, and so on for ever. The whole going-on is strictly determined at the outset, and so we cannot see how it should embrace also the behaviour of living beings including ourselves, who are aware of being able to choose to a large extent the motions of our body by free decision of our mind. If then this mind or soul is itself composed of atoms moving in the same necessitous way, there seems to be no room for ethics or moral behaviour. We are compelled by the laws of physics to do at every moment just exactly the thing we do; what is the good of deliberating whether it is right or wrong? Where is room for the moral law if the natural law overpowers and entirely frustrates it?

The antinomy is as unsolved today as it was twenty-three centuries ago. Still we are able to analyse Democritus' assumption into one very creditable and one very absurd constituent. He admitted

(*a*) that the behaviour of *all* the atoms inside a living body was determined by the physical laws of Nature, and

(*b*) that some of them went to compose what we call mind or soul.

I consider it very much to his credit that he held unswervingly on to (*a*), even though it implies an antinomy, with or without (*b*). Indeed, if you admit (*a*), the motion of your body is predetermined and you fail to account for your sensation that you move it at will, whatever you may think about the mind.

The truly absurd feature is (*b*).

Unfortunately Democritus' successors, Epicurus and his disciples, finding their minds not strong enough to face the antinomy, abandoned the creditable assumption (*a*) and clung to the absurd blunder (*b*).

The difference between the two men, Democritus and Epicurus, was that Democritus was still modestly aware that he knew nothing, while Epicurus was very sure that he knew very little short of everything.

Epicurus added to the system another piece of nonsense conscientiously echoed by all his followers, including, of course, Lucretius Carus. Epicurus was a sensualist of the purest breeding. Where the senses give us conclusive evidence, we must follow it. Where they do not we are allowed to make any reasonable hypothesis to explain what we see. Unfortunately he included in the things about which the senses give us conclusive, indubitable evidence, the size of the sun, the moon and the stars. Speaking in particular about the sun, he argued (*a*) that its circumference is sharp, not blurred, and (*b*) that we feel its heat. He argued further that, if a big terrestrial fire is still near enough for us to discern its contours clearly and to feel anything of its heat, then we also discern its actual size, 'we see it just as big as it is'! Conclusion: the sun (and the moon and the stars) are just as big as we see them, neither bigger nor smaller.

The main nonsense is, of course, the expression 'as big as we see them'. It is astonishing that even modern philologists, when they report on this, are not shaken by this meaningless expression, but only by Epicurus' saying yes to it. He does not distinguish between angular size and linear size—living in Athens nearly three centuries after Thales, who measured the distance of ships by triangulation, as we do.

But let us take his words at the face value. What can he have meant? How big, then, do we see the sun? And how far is it thus away if it is as big as we see it?

The angular size is 1/2 of a degree. From this you easily make out, that if it were 10 miles away, it would have to have a diameter of roughly 1/10 of a mile or 500 feet. I do not think anybody could hold that the sun gives the immediate impression of being even as big as a cathedral. But let us grant him ten or fifteen times the size, which would give a diameter of a mile and a half and a distance of 150 miles. That would mean that when you saw the sun in the morning in Athens on the eastern horizon, it was actually rising from the coast of Asia Minor. Think of it:

Fig. 4.

Did he think it passed horizontally over the Mediterranean? With his ignorance of angular size that is quite possible.

At any rate I think this shows that after Democritus the colours of physics were flown by philosophers who had no real interest in science and who, by the great influence they had as philosophers, wrecked it, in spite of the brilliant specialized work that was going on in Alexandria and elsewhere. It had little influence on the attitude of the population at large, including even such men as Cicero, Seneca or Plutarch.

Let us now return to the historical questions raised at the beginning of this chapter of which I said that they have much more than only historical interest. We are facing here one of the most fascinating cases in the history of ideas. The astonishing point is this. From the lives and writings of Gassendi and Descartes, who introduced atomism into modern science, we know as an actual historical fact that, in doing so, they were fully aware of taking up the theory of the ancient philosophers whose scripts they had diligently studied. Furthermore, and more importantly, all the basic features of the ancient theory have survived in the modern one up to this day, greatly enhanced and widely elaborated but unchanged, if we apply the standard of the natural philosopher, not the myopic perspective of the specialist. On the other hand we know that not a scrap of the wide experimental evidence that a modern physicist adduces in support of those basic features was known either to Democritus or to Gassendi.

Whenever this kind of thing happens one has to envisage two possibilities. The first is that the early thinkers made a lucky guess which later proved to be correct. The second is that the thought pattern in question is not so exclusively based on the recently discovered evidence as the modern thinkers believe, but on the co-operation of much simpler facts, known before, and on the *a priori* structure, or at least the natural inclination, of the human intellect. If the likelihood of the second alternative can be proved, it is of paramount importance. It need not, of course, even if it were certain, induce us to abandon the idea—in our case atomism—as a mere fiction of our mind. But it will give us deeper insight into the origin and nature of our thought picture. These considerations urge us to find out, if possible, how were the ancient philosophers led to their conception of immutable atoms and the void?

For all I know there is no extant evidence to guide us. Today, if we state our own or another person's scientific beliefs, we feel bound to add why we or they hold them or held them. The mere fact that N.N. believed this or that, without motivation, seems uninteresting to us. This was not a very common practice in antiquity. Particularly the so-called doxographoi are usually quite content to tell us e.g. 'Democritus held...'. But it is noteworthy in our present context that Democritus himself regarded his insight as a creation of the intellect. This can be seen from fr. 125, quoted below *in extenso,* and also (fr. 11) from his distinction between two kinds of vehicle for obtaining knowledge, the genuine and the dark. The latter are the senses. They let us down when we try to penetrate into small regions of space. Then the genuine method of obtaining knowledge based on a refined organ of thought comes to our aid. That this refers *inter alia* to the atomic theory is obvious, though in the extant fragment it is not mentioned explicitly.

What then guided his refined organ of thought so as to produce the concept of atoms?

Democritus was intensely interested in geometry, not as a mere enthusiast like Plato; he was a geometer of distinction. The theorem that the volume of a pyramid or a cone is *one-third* of the product of its base and height is to his credit. To him who knows the calculus this is a commonplace, but I have met good mathematicians who had some trouble in remembering the elementary proof they had learnt as schoolboys. Democritus can hardly have arrived at the theorem without using, at one step at least, a substitute for the calculus (so does the schoolboy, viz. the principle of Cavalieri—at least in Austria). Democritus had deep insight into the meaning and into the difficulties of *infinitesimals*. This is proved by an interesting paradox which he obviously met on thinking up that proof. Let a cone be cut in two by a plane parallel to its base; are the two circles, produced by the cut on the two parts (the smaller cone above and the cone-stump below), equal or unequal? If unequal, then, since this would hold for any such cut, the ascending part of the cone's surface would not be smooth but covered with indentations; if you say equal, then for the same reason, would it not mean that all these parallel sections are equal and thus that the cone is a cylinder?

From this and from the extant *titles* of two other scripts ('On a difference of opinion or on the contact of a circle and a sphere'; 'On irrational lines and solids') one gains the impression that he eventually arrived at a clear distinction between, on the one hand, the geometrical concepts of a body, a surface or a line of well-defined properties (as e.g. a pyramid, a square surface or a circular line), and, on the other hand, the more or less imperfect realizations of these concepts by or on a physical body. (Plato, a century later, reckoned the first category among his 'ideas'; nay they were, so I believe, his prototypes thereof; thus the thing got muddled up with metaphysics.)

Now grasp this together with the fact that Democritus not only knew the opinions of the Ionian philosophers, but may be said to have continued their tradition; moreover that the last of them, Anaximenes, as we said in Chapter IV, successfully and in full agreement with our modern views, maintained that all the momentous changes observed in matter were only apparent, actually due to rarefaction and condensation. But is it meaningful to say that the material itself remains unchanged, if actually every bit of it, however small, becomes thinned out or compressed? The *geometer* Democritus was well able to conceive of this 'however small'. The obvious way out is to think of any physical body as actually composed of innumerable small bodies, which remain always unchanged, while rarefaction is produced when they recede from each other, condensation, when they crowd more closely together into a small volume. To allow them to do this, within limits, it is a necessary requirement that the space between them be void, i.e. contain nothing at all. At the same time the integrity of pure geometrical statements could be saved by diverting the paradoxes and challenges from the geometrical concepts to their imperfect physical realizations. The surface of a real cone or, for that matter, of any real body, was actually not smooth, since it was formed by the top layer of atoms and thus was riddled with small holes with prominences between them. It could also be granted to Protagoras (who had put forward challenges of this kind) that a real sphere resting on a real plane had not just one point of contact with it, but a whole small region of 'near' contact. All this would not hamper the exactness of pure geometry. That this was Democritus' view may be inferred from a remark of Simplicius, who tells us that, according to Democritus, his physically indivisible atoms were in a mathematical sense divisible *ad infinitum*.

During the last fifty years we have obtained experimental evidence of the 'real existence of discrete corpuscles'. There is a wide range of most interesting observations that we cannot summarize here and that the atomists at the end of the nineteenth century did not anticipate in their most intemperate dreams. We can see with our own eyes the recorded linear traces of the paths of single elementary particles in the Wilson cloud-chamber and in photographic emulsions. We have instruments (Geiger counters) which respond with an audible click to a single cosmic ray particle that enters the instrument; moreover the latter may be so devised that at every click an ordinary commercial electricity meter is advanced by one, so that it counts the number of particles that have arrived within a given time. Such counts performed by different methods and under varied conditions are in full agreement with each other as well as with the atomic theories developed long before this direct evidence was available. The great atomists from Democritus down to Dalton, Maxwell and Boltzmann would have gone into raptures at these palpable proofs of their belief.

But at the same time modern atomic theory has been plunged into a crisis. There is no doubt that the simple particle theory is too naïve. This is not altogether too astonishing, from the above speculations about its origin. If these are correct, then atomism was forged as a weapon to overcome the difficulties of the mathematical continuum, of which, as we have seen, Democritus was fully aware. To him atomism was a means for bridging the gulf between the real bodies of physics and the idealized geometrical shapes of pure mathematics. But not only to Democritus. In a way atomism has performed this task all through its long history, the task of facilitating our thinking about palpable bodies. A piece of matter is resolved in our thought into an innumerably great, yet finite number of constituents. We can imagine our *counting* them, while we are unable to tell the number of points on a straight line of 1 cm. length. We can *count,* in our thought, the number of mutual impacts within a given time. When hydrogen and chlorine unite to form hydrochloric acid, we can, in our mind, pair off the atoms of the two kinds and think that every pair unites to form a new little body, a molecule of the compound. This counting, this pairing off, this whole manner of thinking has played a prominent role in discovering the most important physical theorems. It would seem impossible under the aspect that matter is a continuous structureless jelly. Thus atomism has proved infinitely fruitful. Yet the more one thinks of it, the less can one help wondering to what extent it is a *true* theory. Is it really founded exclusively on the actual objective structure of 'the real world around us'? Is it not in an important way conditioned by the nature of human understanding—what Kant would have called '*a priori*'? It behoves us, so I believe, to preserve an extremely open mind towards the palpable proofs of the existence of individual single particles, without detriment to our deep admiration for the genius of those experimenters who have furnished us with this wealth of knowledge. They are increasing it from day to day and are thereby helping to turn the scales with respect to the sad fact, that our theoretical understanding thereof is, I venture to say, diminishing at almost the same rate.

Let me conclude this chapter by quoting some agnostic and sceptic fragments of Democritus, which have impressed me most. The translations follow Cyril Bailey.

(D. fr. 6) A man must learn on this principle that he is far removed from the truth.

(D. fr. 7) We know nothing truly about anything, but for each of us his opinion is an influx (i.e. is conveyed to him by an influx of 'idols'[1] from without).

(D. fr. 8) To learn truly what each thing is, is a matter of uncertainty.

(D. fr. 9) In truth we know nothing unerringly, but only as it changes according to the disposition of our body, and of the things that enter into it and impinge on it.

(D. fr. 117) We know nothing truly, for the truth lies hidden in the depth.

And now the famous dialogue between the intellect and the senses:

(D. fr. 125) (*Intellect:*) Sweet is by convention and bitter by convention, hot by convention, cold by convention, colour by convention; in truth there are but atoms and the void.

(*The Senses:*) Wretched mind, from us you are taking the evidence by which you would overthrow us? Your victory is your own fall.

[1] Gk. εἴδωλον, picture.

CHAPTER VII

WHAT ARE THE SPECIAL FEATURES?

Let me now, at last, approach the answer to the question which was put at the outset.

Remember the lines of Burnet's preface—that *science* is a Greek invention; that science has never existed except among peoples who came under Greek influence. Later in the same book he says: 'The founder of the Milesian School and therefore [!] the first man of science was Thales.'[2] Gomperz says (I quoted him extensively) that our whole modern way of thinking is based on Greek thinking; it is therefore something special, something that has grown historically over many centuries, *not* the general, the only possible way of thinking about Nature. He sets much store on our becoming aware of this, of recognizing the peculiarities as such, possibly freeing us from their wellnigh irresistible spell.

What are they then? What are the peculiar, special traits of our scientific world-picture?

About one of these fundamental features there can be no doubt. It is the hypothesis that *the display of Nature can be understood.* I have touched on this point repeatedly. It is the non-spiritistic, the non-superstitious, the non-magical outlook. A lot more could be said about it. One would in this context have to discuss the questions: what does comprehensibility really

[2] *Early Greek Philosophy*, p. 40.

mean, and in what sense, if any, does science give explanations? David Hume's (1711–76) great discovery that the relation between cause and effect is not directly observable and enunciates nothing but the regular succession—this fundamental epistemological discovery has led the great physicists, Gustav Kirchhof (1824–87) and Ernst Mach (1838–1916), and others to maintain that natural science does not vouchsafe any explanations, that it aims only at, and is unable to attain to anything but, a complete and (Mach) economical description of the observed facts. This view, in the more elaborate form of philosophical positivism, has been enthusiastically embraced by modern physicists. It has great consistency; it is very difficult, if not impossible, to refute, rather like solipsism, but is very much more reasonable than the latter. Though the positivist view ostensibly contradicts the 'understandability of Nature', it is certainly not a return to the superstitious and magical outlook of yore; quite the contrary, from physics it expels the notion of force, the most dangerous relic of animism in this science. It is a salutary antidote against the rashness with which scientists are prone to believe that they have understood a phenomenon, when they have really only grasped the facts by describing them. Yet even from the positivists' point of view one ought not, so I believe, to declare that science conveys no understanding. For even if it be true (as they maintain) that in principle we only observe and register facts and put them into a convenient mnemotechnical arrangement, there are factual relations between our findings in the various, widely distant domains of knowledge, and again between them and the most fundamental general notions (as the natural integers 1, 2, 3, 4, ...), relations so striking and interesting, that for our eventual grasping and registering them the term 'understanding' seems very appropriate. The most outstanding examples, to my mind, are the mechanical theory of heat, which amounts to a reduction to pure numbers; and similarly I would call Darwin's theory of evolution an instance of our gaining true insight. The same can be said about genetics, based on the discoveries of Mendel and de Vries, while in physics quantum theory has reached a promising outlook, but has not yet attained to full comprehensibility, though it is successful and helpful in many ways, even in genetics and biology in general.

There is, however, so I believe, a second feature, much less clearly and openly displayed, but of equally fundamental importance. It is this, that science in its attempt to describe and understand Nature simplifies this very difficult problem. The scientist subconsciously, almost inadvertently, simplifies his problem of understanding Nature by disregarding or cutting out of the picture to be constructed, himself, his own personality, the subject of cognizance.

Inadvertently the thinker steps back into the role of an external observer. This facilitates the task very much. But it leaves gaps, enormous lacunae, leads to paradoxes and antinomies whenever, unaware of this initial renunciation, one tries to find oneself in the picture or to put oneself, one's own thinking and sensing mind, back into the picture.

This momentous step—cutting out oneself, stepping back into the position of an observer who has nothing to do with the whole performance—has received other names, making it appear quite harmless, natural, inevitable. It might be called just objectivation, looking upon the world as an object. The moment you do that, you have virtually ruled yourself out. A frequently used expression is 'the hypothesis of a real world around us' (*Hypothese der realen Aussenwelt*). Why, only a fool would forgo it! Quite right, only a fool. None the less it is a definite trait, a definite feature of our way of understanding Nature—and it has consequences.

The clearest vestiges of this idea that I could find in ancient Greek writing are those fragments of Heraclitus that we have been discussing and analysing just before. For it is this 'world in common', this *ξυνόν* or *κοινόν* of Heraclitus, that we are constructing; we are hypostatizing the world as an object, making the assumption of a real world around us—as the most popular phrase runs—made up of the overlapping parts of our several consciousnesses. And in doing so, everyone willy-nilly takes himself—the subject of cognizance, the thing that says 'cogito ergo sum'—out of the world, removes himself from it into the position of an external observer, who does not himself belong to the party. The 'sum' becomes 'est'.

Is that really so, must it be so, and why is it so? For we are not aware of it. I'll say presently why we are not aware of it. First let me say why it is so.

Well, the 'real world around us' and 'we ourselves', i.e. our minds, are made up of the same building material, the two consist of the same bricks, as it were, only arranged in a different order—sense perceptions, memory images, imagination, thought. It needs, of course, some reflexion, but one easily falls in with the fact that matter is composed of these elements and nothing else. Moreover, imagination and thought take an increasingly important part (as against crude sense-perception), as science, knowledge of nature, progresses.

What happens is this. We can think of these—let me call them *elements*—either as constituting mind, everyone's own mind, or as constituting the material world. But we cannot, or can only with great difficulty, think both things at the same time. To get from the mind-aspect to the matter-aspect or vice versa, we have, as it were, to take the elements asunder and to put them together again in an entirely different order. For example—it is not easy to give examples, but I'll try—my mind at this moment is constituted by all I sense around me: my own body, you all sitting in front of me and very kindly listening to me, the *aide-mémoire* in front of me, and, above all, the ideas I wish to explain to you, the suitable framing of them into words. But now envisage any one of the material objects around us, for example my arm and hand. As a material object it is composed, not only of my own direct sensations of it, but also of the imagined sensations I would have in turning it round, moving it, looking at it from all different angles; in addition it is composed of the perceptions I imagine you to have of it, and also, if you think of it purely scientifically, of all you could verify and would actually find, if you took it and dissected it, to convince yourself of its intrinsic nature and composition. And so on. There is no end to enumerating all the potential percepts and sensations on my and on your side that are included in my speaking of this arm as of an objective feature of the 'real world around us'.

The following simile is not very good, but it is the best I can think of: a child is given an elaborate box of bricks of various sizes and shapes and colours. It can build from them a house, or a tower, or a church, or the Chinese wall, etc. But it cannot build two of them at the same time, because it is, at least partly, the same bricks it needs in every case.

This is the reason why I believe it to be true that I actually do cut out my mind when I construct the real world around me. And I am not aware of this cutting out. And then I am very astonished that the scientific picture of the real world around me is very deficient. It gives a lot of factual information, puts all our experience in a magnificently consistent order, but it is ghastly silent about all and sundry that is really near to our heart, that really matters to us. It cannot tell us a word about red and blue, bitter and sweet, physical pain and physical delight; it knows nothing of beautiful and ugly, good or bad, God and eternity. Science sometimes pretends to answer questions in these domains, but the answers are very often so silly that we are not inclined to take them seriously.

So, in brief, we do not belong to this material world that science constructs for us. We are not in it, we are outside. We are only spectators. The reason why we believe that we are in it, that we belong to the picture, is that our bodies are in the picture. Our bodies belong to it. Not only my own body, but those of my friends, also of my dog and cat and horse, and of all the other people and animals. And this is my only means of communicating with them.

Moreover, my body is implied in quite a few of the more interesting changes—movements, etc.—that go on in this material world, and is implied in such a way that I feel myself partly the author of these goings-on. But then comes the impasse, this very embarrassing discovery of science, that I am not needed as an author. Within the scientific world-picture all these happenings take care of themselves, they are amply accounted for by direct energetic interplay. Even the human body's movements 'are its own' as Sherrington put it. The scientific world-picture vouchsafes a very complete understanding of all that happens—it makes it just a little too understandable. It allows you to imagine the total display as that of a mechanical clock-work, which for all that science knows could go on just the same as it does, without there being consciousness, will, endeavour, pain and delight and responsibility connected with it—though they actually are. And the reason for this disconcerting situation is just this, that, for the purpose of constructing the picture of the external world, we have used the greatly simplifying device of cutting our own personality out, removing it; hence it is gone, it has evaporated, it is ostensibly not needed.

In particular, and most importantly, this is the reason why the scientific world-view contains of itself no ethical values, no aesthetical values, not a word about our own ultimate scope or destination, and no God, if you please. Whence came I, whither go I?

Science cannot tell us a word about why music delights us, of why and how an old song can move us to tears.

Science, we believe, can, in principle, describe in full detail all that happens in the latter case in our sensorium and 'motorium' from the moment the waves of compression and dilation reach our ear to the moment when certain glands secrete a salty fluid that emerges from our eyes. But of the feelings of delight and sorrow that accompany the process science is completely ignorant—and therefore reticent.

Science is reticent too when it is a question of the great Unity—the One of Parmenides—of which we all somehow form part, to which we belong. The most popular name for it in our time is God—with a capital 'G'. Science is, very usually, branded as being atheistic. After what we said, this is not astonishing. If its world-picture does not even contain blue, yellow, bitter, sweet—beauty, delight and sorrow—, if personality is cut out of it by agreement, how should it contain the most sublime idea that presents itself to human mind?

The world is big and great and beautiful. My scientific knowledge of the events in it comprises hundreds of millions of years. Yet in another way it is ostensibly contained in a poor seventy or eighty or ninety years granted to me—a tiny spot in immeasurable time, nay even in the finite millions and milliards of years that I have learnt to measure and to assess. Whence come I and whither go I? That is the great unfathomable question, the same for every one of us. Science has no answer to it. Yet science represents the level best we have been able to ascertain in the way of safe and incontrovertible knowledge.

However, our life as something like human beings has lasted, at the most, only about half a million years. From all that we know, we may anticipate, even on this particular globe, quite a few million years to come. And from all this we feel that any thought we attain to during this time will not have been thought in vain.

BIBLIOGRAPHY

BAILEY, CYRIL. *The Greek Atomists and Epicurus.* Oxford University Press, 1928.

—— *Epicurus.* Oxford University Press, 1926 (extant texts with translation and commentary).

—— *Translation of Lucretius' De rerum natura* (with introduction and notes). Oxford University Press, 1936.

BURNET, JOHN. *Early Greek Philosophy.* London: A. and C. Black, 1930 (4th ed.).

—— *Greek Philosophy, Thales to Plato*, London: Macmillan and Co., 1932.

DIELS, HERMANN. *Die Fragmente der Vorsokratiker.* Berlin: Weidmann, 1903 (1st ed.).

FARRINGTON, BENJAMIN. *Science and Politics in the Ancient World.* London: Allen and Unwin, 1939.

—— *Greek Science*, I (Thales to Aristotle); II (Theophrastus to Galen). Pelican.

GOMPERZ, THEODOR. *Griechische Denker.* Leipzig: Veit and Comp., 1911.

HEATH, SIR THOMAS L. *Greek Astronomy.* London: J. M. Dent and Sons, 1932.

—— *A Manual of Greek Mathematics.* Oxford University Press, 1931.

HEIBERG, J. L. *Mathematics and Physical Science in Classical Antiquity.* Oxford University Press, 1922.

MACH, ERNST. *Populärwissenschaftliche Vorlesungen.* Leipzig: J. A. Barth, 1903.

MUNRO, H. A. *Titus Lucretius Carus, De rerum natura.* Cambridge, Deighton, Bell and Co., 1889.

RUSSELL, BERTRAND. *History of Western Philosophy.* London: Allen and Unwin, 1946.

SCHRÖDINGER, E. 'Die Besonderheit des Weltbilds der Naturwissenschaft'. *Acta Physica Austriaca* **1**, 201, 1948.

SHERRINGTON, SIR CHARLES. *Man on his Nature.* Cambridge University Press, 1940 (1st ed.).

WINDELBAND, WILHELM. *Geschichte der Philosophie.* Tübingen und Leipzig: J. C. B. Mohr, 1903.

SCIENCE AND HUMANISM

o o

Physics in our time

To
my companion
through thirty years

PREFACE

These are four public lectures which were delivered under the auspices of the Dublin Institute for Advanced Studies at University College Dublin in February, 1950 under the title 'Science as a Constituent of Humanism'. Neither this nor the abbreviated title chosen here adequately covers the whole, but rather the first sections only. In the remaining part, from p. 11 onward, I intend to depict the present situation in physics as it has gradually developed in the current century; to depict it from the point of view expressed in the title and in the earlier part, thus giving, as it were, an example of how I am looking on scientific effort: as forming part of man's endeavour to grasp the human situation.

My thanks are due to the Cambridge University Press for the rapid production of this booklet and to Miss Mary Houston from the Dublin Institute for designing the figures and reading the proofs.

March 1951 E.S.

SCIENCE AND HUMANISM

THE SPIRITUAL BEARING OF SCIENCE ON LIFE

What is the value of scientific research? Everybody knows that in our days more than ever before a man or a woman who wishes to make a genuine contribution to the advancement of science has to specialize: which means to intensify one's endeavour to learn all that is known within a certain narrow domain and then to try and increase this knowledge by one's own work—by studies, experiments, and thinking. Being engaged in such specialized activity one naturally at times stops to think what it is good for. Has the promotion of knowledge within a narrow domain any value in itself? Has the sum total of achievements in all the several branches of *one* science—say of physics, or chemistry, or botany, or zoology—any value in itself—or perhaps the sum total of the achievements of all the sciences together—and *what* value has it?

A great many people, particularly those not deeply interested in science, are inclined to answer this question by pointing to the practical consequences of scientific achievements in transforming technology, industry, engineering, etc., in fact in changing our whole way of life beyond recognition in the course of less than two centuries, with further and even more rapid changes to be expected in the time to come.

Few scientists will agree with this utilitarian appraisal of their endeavour. Questions of values are, of course, the most delicate ones; it is hardly possible to offer incontrovertible arguments. But let me give you the three principal ones by which I should try to oppose this opinion.

Firstly, I consider natural science to be very much on the same line as the other kinds of learning—or *Wissenschaft*, to use the German expression—cultivated at our universities and other centres for the advancement of knowledge. Consider the study or research in history or languages, philosophy, geography—or history of music, painting, sculpture, architecture—or in archaeology and pre-history; nobody would like to associate with these activities, as their principal aim, the practical improvement of the conditions of human society, although improvement does result from them quite frequently. I cannot see that science has, in this respect, a different standing.

On the other hand (and this is my second argument), there are natural sciences which have obviously no practical bearing at all on the life of the human society: astrophysics, cosmology, and some branches of geophysics. Take, for instance, seismology. We know enough about earthquakes to know that there is very little chance of foretelling them, in the way of warning people to leave their houses, as we warn trawlers to return when a storm is drawing near. All that seismology could do is to warn prospective settlers of certain danger zones; but those, I am afraid, are mostly known by sad experience without the aid of science, yet they are often densely populated, the need for fertile soil being more pressing.

Thirdly, I consider it extremely doubtful whether the happiness of the human race has been enhanced by the technical and industrial developments that followed in the wake of rapidly progressing natural science. I cannot here enter into details, and I will not speak of the future development—the surface of the earth getting infected with artificial radioactivity, with the gruesome consequences for our race, depicted by Aldous Huxley in his horribly interesting recent novel (*Ape and Essence*). But consider only the 'marvellous reduction of size' of the world by the fantastic modern means of traffic. All distances have been reduced to almost nothing, when measured not in miles but in hours of *quickest* transport. But when measured in the costs of even the *cheapest* transport they have been doubled or trebled even in the last 10 or 20 years. The result is that many families and groups of close friends have been scattered over the globe as never before. In many cases they are not rich enough ever to meet again, in others they do so under terrible sacrifices for a short time ending in a heart-rending farewell. Does that make for human happiness? These are a few striking examples; one could enlarge on the topic for hours.

But let us turn to less gloomy aspects of human activities. You may ask—you are bound to ask me now: What, then, is in your opinion the value of natural science? I answer: Its scope, aim and value is the same as that of any other branch of human knowledge. Nay, none of them alone, only the union of all of them, has any scope or value at all, and that is simply enough described: it is to obey the command of the Delphic deity, Γνῶθι σεαυτόν, get to know yourself. Or, to put it in the brief, impressive rhetoric of Plotinus (*Enn.* VI, 4, 14): ἡμεῖς δέ, τίνες δὲ ἡμεῖς; 'And we, who are we anyhow?' He continues: 'Perhaps we were *there* already before this creation came into existence, human beings of another type, or even some sort of gods, pure souls and mind united with the whole universe, parts of the intelligible world, not separated and cut off, but at one with the whole.'

I am born into an environment—I know not whence I came nor whither I go nor who I am. This is my situation as yours, every single one of you. The fact that everyone always was in this same situation, and always will be, tells me nothing. Our burning question as to the whence and whither —all we can ourselves observe about it is the present environment. That is why we are eager to find out about it as much as we can. That is science, learning, knowledge, that is the true source of every spiritual endeavour of man. We try to find out as much as we can about the spatial and temporal surrounding of the place in which we find ourselves put by birth.

And as we try, we delight in it, we find it extremely interesting. (May *that* not be the end for which we are there?)

It seems plain and self-evident, yet it needs to be said: the isolated knowledge obtained by a group of specialists in a narrow field has in itself no value whatsoever, but only in its synthesis with all the rest of knowledge and only inasmuch as it really contributes in this synthesis something toward answering the demand τίνες δὲ ἡμεῖς; ('who are we?')

José Ortega y Gasset, the great Spanish philosopher, who is now after many years of exile back in Madrid (though he is, I believe, just as little a fascist as a *sozialdemokrat*, but just an ordinary reasonable person), published in the twenties of this century a series of articles, which were later collected in a delightful volume under the title of *La rebelión de las masas*—the rebellion of the masses. It has, by the way, nothing to do with social or other revolutions, the *rebelión* is meant purely metaphorically. The Age of Machinery has resulted in sending the numbers of the populations and the volume of their needs up to enormous heights, unprecedented and unforeseeable. The daily life of every one of us becomes more and more entangled with the necessity of coping with these numbers. Whatever we need or desire, a loaf of bread or a pound of butter, a bus-lift or a theatre-ticket, a quiet holiday resort or the permit to travel abroad, a room to live in or a job to live on . . . there are always many, many others having the same need or desire. The new situations and developments that have turned up as the result of this unparalleled soaring of the numbers form the subject of Ortega's book.

It contains extremely interesting observations. Just to give you an example—though it does not concern us at the moment—one chapter-heading reads *El major peligro, el estado*: the greatest danger—the state. He there declares the increasing power of the state in curtailing individual freedom—under the pretext of protecting us, but far beyond necessity—to be the greatest danger to the future development of culture (*kultur*). But the chapter I wish to speak of here is the preceding one; it is entitled *La barbarie del 'especialismo'*: the barbarism of specialization. At first sight it seems paradoxical and it may shock you. He makes bold to picture the specialized scientist as the typical representative of the brute ignorant rabble—the *hombre masa* (mass-man)—who endanger the survival of true civilization. I can only pick out a few passages from the delightful description he gives of this 'type of scientist without precedent in history'.

> He is a person who, of all the things that a truly educated person ought to know of, is familiar only with one particular science, nay even of this science only that small portion is known to him in which he himself is engaged in research. He reaches the point where he proclaims it a virtue not to take any notice of all that remains outside the narrow domain he himself cultivates, and denounces as *dilettantist* the curiosity that aims at the synthesis of all knowledge.

It comes to pass that he, secluded in the narrowness of his field of vision, actually succeeds in discovering new facts and in promoting his science (which he hardly knows) and promoting along with it the integrated human thought—which he with full determination ignores. How has anything like this been possible, and how does it continue to be possible? For we must strongly underline the inordinateness of this undeniable fact: experimental science has been advanced to a considerable extent by the work of fabulously mediocre and even less than mediocre persons.

I shall not continue the quotation, but I strongly recommend you to get hold of the book and continue for yourself. In the twenty-odd years that have passed since the first publication, I have noticed very promising traces of opposition to the deplorable state of affairs denounced by Ortega. Not that we can avoid specialization altogether; that is impossible if we want to get on. Yet the awareness that specialization is not a virtue but an unavoidable evil is gaining ground, the awareness that all specialized research has real value only in the context of the integrated totality of knowledge. The voices become fainter and fainter that accuse a man of dilettantism who dares to think and speak and write on topics that require more than the special training for which he is 'licensed' or 'qualified'. And any loud barking at such attempts comes from very special quarters of two types—either very scientific or very unscientific quarters—and the reasons for the barking are in both cases translucent.

In an article on 'The German Universities' (published on 11 December 1949 in *The Observer*)

Robert Birley, Headmaster of Eton, quoted some lines from the report of the Commission for University Reform in Germany—quoted them very emphatically, an emphasis that I fully endorse. The following is said in this report:

> Each lecturer in a technical university should possess the following abilities:
>
> (a) To see the limits of his subject matter. In his teaching to make the students aware of these limits, and to show them that beyond these limits forces come into play which are no longer entirely rational, but arise out of life and human society itself.
>
> (b) To show in every subject the way that leads beyond its own narrow confines to broader horizons of its own. Etc.

I won't say that these formulations are peculiarly original, but who would expect originality of a committee or commission or board or that sort of thing? —mankind *en masse* is always very commonplace. Yet one is glad and thankful to find this sort of attitude prevailing. The only criticism—if it be a criticism—is that one can see no earthly reason why these demands should be restricted to the teachers at *technical* universities in *Germany*. I believe they apply to *any* teacher at *any* university, nay, at any school in the world; I should formulate the demand thus:

Never lose sight of the role your particular subject has within the great performance of the tragicomedy of human life; keep in touch with life—not so much with practical life as with the ideal background of life, which is ever so much more important; and, *Keep life in touch with you*. If you cannot

—in the long run—tell everyone what you have been doing, your doing has been worthless.

THE PRACTICAL ACHIEVEMENTS OF SCIENCE TENDING TO OBLITERATE ITS TRUE IMPORT

I regard the public lectures which the statute of the Institute prescribes for us to deliver every year as one of the means for establishing and keeping up this contact in our small domain. Indeed I consider this to be their exclusive scope. The task is not very easy. For one has to have some kind of background to start from, and, as you know, scientific education is fabulously neglected, not only in this or that country—though, indeed, in some more than in others. This is an evil that is inherited, passed on from generation to generation. The majority of educated persons are not interested in science, and are not aware that scientific knowledge forms part of the idealistic background of human life. Many believe—in their complete ignorance of what science really is—that it has mainly the ancillary task of inventing new machinery, or helping to invent it, for improving our conditions of life. They are prepared to leave this task to the specialists, as they leave the repairing of their pipes to the plumber. If persons with this outlook decide upon the curriculum of our children, the result is necessarily such as I have just described it.

There are, of course, historical reasons why this attitude still prevails. The bearing of science on the idealistic background of life has always been great—apart perhaps from the Dark Ages, when science practically did not exist in Europe. But it must be confessed that there has been a lull also in more recent times, which could easily deceive one into under-rating the idealistic task of science. I place the lull about in the second half of the nineteenth century. This was a period of enormous explosion-like development of science, and along with it of a fabulous, explosion-like development of industry and engineering which had such a tremendous influence on the material features of human life that most people forgot any other connections. Nay, worse than that! The fabulous *material* development led to a *materialistic* outlook, allegedly derived from the new scientific discoveries. These occurrences have, I think, contributed to the deliberate neglect of science in many quarters during the half century that followed—the one that is just drawing to a close. For there always is a certain time-lag between the views held by learned men and the views held by the general public about the views of those learned men. I do not think that fifty years is an excessive estimate for the average length of that time-lag.

Be that as it may, the fifty years that have just gone by—the first half of the twentieth century—have seen a development of science in general, and of physics in particular, unsurpassed in transforming our Western outlook on what has often been called the Human Situation. I have little doubt that it will take another fifty years or so before the educated section of the general public will have become aware of this change. Of course, I am not so much of an idealistic dreamer as to hope substantially to accelerate this process by a couple of public lectures. But, on the other hand, this process of *assimilation* is not automatic. *We have to labour for it.* In this labour I take my share, trusting that others will take theirs. It is part of our task in life.

A RADICAL CHANGE IN OUR IDEAS OF MATTER

We shall now, at last, come down to some special topics. What I have said hitherto may seem pretty long, if you consider it a mere introduction. But I hope it is of some interest in itself—and I could not avoid it. I had to make clear the situation. None of the new discoveries about which I may tell you is frightfully exciting in itself. What *is* exciting, novel, revolutionary, is the general attitude we are compelled to adopt on any attempt to synthesize them all.

Let us go *in medias res.* There is the problem of *matter.* What *is* matter? How are we to picture *matter* in our *mind*?

The first form of the question is ludicrous. (How should we say *what* matter *is*—or, if it comes to that, *what electricity is*—both being phenomena given to us once only?) The second form already betrays the whole change of attitude: matter is an image in our mind—mind is thus prior to matter (notwithstanding the strange empirical dependence of my mental processes on the physical data of a certain portion of matter, viz. my brain).

During the second half of the nineteenth century matter seemed to be the permanent thing to which we could cling. *There* was a piece of matter that had never been created (as far as the physicist knew) and could never be destroyed! You could hold on to it and feel that it would not dwindle away under your fingers.

Moreover this matter, the physicist asserted, was with regard to its demeanour, its motion, subject to rigid laws—every bit of it was. It moved according to the forces which neighbouring parts of matter, according to their relative situations, exerted on it. You could *foretell* the behaviour, it was rigidly determined in all the future by the initial conditions.

This was all quite pleasing, anyhow in physical science, in so far as external inanimate matter comes into play. When applied to the matter that constitutes our own body or the bodies of our friends, or even that of our cat or our dog, a well-known difficulty arises with regard to the apparent freedom of living beings to move their limbs at their own will. We shall enter on this question later (see p. 58 ff.) At the moment I wish to try and explain the radical change in our ideas about matter that has taken place in the course of the last half-century. It came about gradually, inadvertently, without anybody aiming at such a change. We believed we moved still within the old 'materialistic' frame of ideas, when it turned out that we had left it.

Our conceptions of matter have turned out to be 'much less materialistic' than they were in the second half of the nineteenth century. They are still very imperfect, very hazy, they lack clearness in various respects; but this can be said, that matter has ceased to be the simple palpable coarse thing in space that you can follow as it moves along, every bit of it, and ascertain the precise laws governing its motion.

Matter is constituted of particles, separated by comparatively large distances; it is embedded in empty space. This notion goes back to Leucippus and Democritus, who lived in Abdera in the fifth century B.C. This conception of particles and empty space (ἄτομοι καὶ κενόν) is retained today (with a modification that is just the thing I wish to explain now)—and not only that, there is complete historical continuity; that is to say, whenever the idea was taken up again it was in full awareness of the fact that one was taking up the concepts of the ancient philosophers. Moreover it experienced the greatest thinkable triumphs in actual experiment, such as the ancient philosophers would hardly have hoped for in their boldest dreams. For instance, O. Stern succeeded in determining the distribution of velocities among the atoms in a jet of silver vapour by the simplest and most natural means, of which figure 1 gives a rough schematical sketch. The outer circle (carrying the letters A, B, C) represents the cross-section of a closed cylindrical box, exhausted to perfect vacuum. The point S marks the cross-section of an incandescent silver wire, which extends along the axis of the cylinder and continually evaporates silver atoms, that fly along straight lines, roughly speaking, in radial directions. However, the cylindrical shield Sh (smaller circle), disposed concentrically around S, lets them pass only at the opening O,

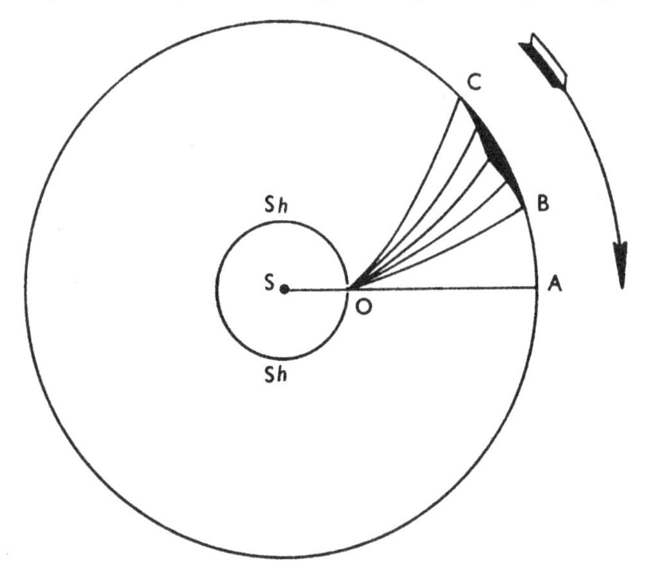

Figure 1

which represents a narrow slit parallel to the wire S. Without anything more, they pass on straight to A, where they are caught and, after a time, form a precipitate in the form of a narrow black line (parallel to the wire S and the slit O). But in Stern's experiment *the whole apparatus* is rotated, as on a potter's wheel, with high speed around the axis S (the sense of the rotation shown by the arrow). This has the effect that the flying atoms—which are, of course, *not* affected by the rotation—are not precipitated at A but at points 'behind' A, the *farther* behind, the *slower* they are, because they allow the collecting surface to turn through a *bigger* angle before they reach it. Thus the slowest atoms form a line at C, the quickest at B. After a time one obtains a broad band whose cross-section is schematically indicated in our figure. By measuring its varying thickness and taking into account the dimensions of the apparatus and its speed of rotation, one can determine the actual velocity of the atoms, more particularly the relative numbers of atoms flying with various velocities—the so-called velocity distribution. I must still explain the fan-like spreading of the paths of the atoms and their *curvature* shown in the figure, both in apparent contradiction to what I said about the flying atoms *not* being affected by the rotation of the apparatus. I have taken the liberty to draw these lines though they are *not* the 'actual' paths of the atoms, but what their paths would appear to be to an observer sharing the rotation of the apparatus (just as we share the rotation of the earth). It is essential to make oneself clear that these 'relative paths' remain the same during the rotation. Hence we may continue the rotation as long as we please, for a substantial deposit to develop.

These important experiments confirmed quantitatively Maxwell's theory of gases, many years after this theory had been expounded. Today they have been eclipsed, and all but forgotten, by far more impressive investigations.

The effect of a single fast particle can be observed as it impinges on a fluorescent screen and causes a faint flash of light, a scintillation. (If you have a watch with luminous figures, take it into a dark room and observe it with a moderately strong magnifying glass: you will then observe the scintillations caused by the impact of single He-ions, α-particles, as they are called in this context.) In a Wilson cloud chamber you can observe the paths of single particles, α-particles, electrons, mesons, . . . , their traces can be photographed and you can determine their curvature in a magnetic field; cosmic ray particles passing through a photographic emulsion produce nuclear disintegrations there, and both the primary and the secondary particles (if they are charged, as they usually are) trace their paths in the emulsion, so that the paths become visible when the plate is developed by the ordinary photographic procedure. I could give you more examples (but these will suffice) of the very direct way in which the old hypothesis of the particle structure of matter has been confirmed far beyond the keenest expectation of previous centuries.

Still less expected is the modification which our ideas about the nature of all these particles underwent during the same time—had to undergo willy-nilly—in consequence of other experiments and of theoretical considerations.

Democritus and all who followed on his path up

to the end of the nineteenth century, though they had never traced the effect of an individual atom (and probably did not hope ever to be able to), were yet convinced that the atoms *are* individuals, identifiable, small bodies just like the coarse palpable objects in our environment. It seems almost ludicrous that precisely in the same years or decades which let us succeed in tracing single, individual atoms and particles, and that in various ways, we have yet been compelled to dismiss the idea that such a particle is an individual entity which in principle retains its 'sameness' for ever. Quite the contrary, we are now obliged to assert that the ultimate constituents of matter have no 'sameness' at all. When you observe a particle of a certain type, say an electron, now and here, this is to be regarded in principle as an *isolated event*. Even if you do observe a similar particle a very short time later at a spot very near to the first, and even if you have every reason to assume a *causal connection* between the first and the second observation, there is no true, unambiguous meaning in the assertion that it is *the same* particle you have observed in the two cases. The circumstances may be such that they render it highly convenient and desirable to express oneself so, but it is only an abbreviation of speech; for there are other cases where the 'sameness' becomes entirely meaningless; and there is no sharp boundary, no clear-cut distinction between them, there is a gradual transition over intermediate cases. And I beg to emphasize this and I beg you to believe it: It is not a question of our being able to ascertain the identity in some instances and not being able to do so in others. It is beyond doubt that the question of 'sameness', of identity, really and truly has no meaning.

FORM, NOT SUBSTANCE, THE FUNDAMENTAL CONCEPT

The situation is rather disconcerting. You will ask: What are these particles then, if they are not individuals? And you may point to another kind of gradual transition, namely that between an ultimate particle and a palpable body in our environment, to which we do attribute individual sameness. A number of particles constitute an atom. Several atoms go to compose a molecule. Molecules there are of various sizes, small ones and big ones, but without there being any limit beyond which we call it a big molecule. In fact there is no upper limit to the size of a molecule, it may contain hundreds of thousands of atoms. It may be a virus or a gene, visible under the microscope. Finally we may observe that any palpable object in our environment is composed of molecules, which are composed of atoms, which are composed of ultimate particles . . . and if the latter lack individuality, how does, say, my wrist-watch come by individuality? Where is the limit? How does individuality arise at all in objects composed of non-individuals?

It is useful to consider this question in some detail, for it will give us the clue to what a particle or an atom really is—what there is permanent in it in spite of its lack of individuality. On my writing-table at home I have an iron letter-weight in the shape of a Great Dane, lying with his paws crossed in front of him. I have known it for many years. I saw it on my father's writing-desk when my nose would hardly reach up to it. Many years later, when my father died, I took the Great Dane, because I liked it, and I used it. It accompanied me to many places, until it stayed behind in Graz in 1938, when I had to leave in something of a hurry. But a friend of mine knew that I liked it so she took it and kept it for me. And three years ago, when my wife visited Austria, she brought it to me, and there it is again on my desk.

I am quite sure it is the same dog, the dog that I first saw more than fifty years ago on my father's desk. But *why* am I sure of it? That is quite obvious. It is clearly the peculiar *form* or *shape* (German: *Gestalt*) that raises the identity beyond doubt, not the material content. Had the material been melted and cast into the shape of a man, the identity would be much more difficult to establish. And what is more: even if the material identity were established beyond doubt, it would be of very restricted interest. I should probably not care very much about the identity or not of that mass of iron, and should declare that my souvenir had been destroyed.

I consider this a good analogy, and perhaps more than an analogy, for pointing out what the particles or atoms really are. For we can see in this example as in many others how in palpable bodies, composed of many atoms, individuality arises out of the structure of their composition, out of shape or form, or organization, as we might call it in other cases. The identity of the *material*, if there is any, plays a subordinate role. You may see this particularly well in cases when you speak of 'sameness' though the material has definitely changed. A man returns after twenty years of absence to the cottage where he spent his childhood. He is profoundly moved by finding the place unchanged. The *same* little stream flows through the *same* meadows, with the cornflowers and poppies and willow trees he knew so well, the white-and-brown cows and the ducks on the pond, as before, and the collie dog coming forth with a friendly bark and wagging his tail to him. And so on. The shape and the organization of the whole place have remained the same, in spite of the entire 'change of material' in many of the items mentioned, including, by the way, our traveller's own bodily self! Indeed, the body he wore as a child has in the most literal sense 'gone with the wind'. Gone, and yet not gone. For, if I am allowed to continue my novelistic snapshot, our traveller will now settle down, marry, and have a small son, who is the very image of his father as old photographs show him at the same tender age.

Let us now return to our ultimate particles and to small organisations of particles as atoms or small molecules. The *old* idea about them was that *their* individuality was based on the identity of matter in them. This seems to be a gratuitous and almost mystical addition that is in sharp contrast to what we have just found to constitute the individuality of macroscopic bodies, which is quite independent of such a crude materialistic hypothesis and does not need its support. The *new* idea is that what is permanent in these ultimate particles or small aggregates is their shape and organization. The habit of everyday language deceives us and seems to require, whenever we hear the word 'shape' or 'form' pronounced, that it must be the shape or form of *something*, that a material substratum is required to take on a shape. Scientifically this habit goes back to Aristotle, his *causa materialis* and *causa formalis*. But when you come to the ultimate particles constituting matter, there seems to be no point in thinking of them again as consisting of some material. They are, as it were, *pure shape*, nothing but shape; what turns up again and again in successive observations is this shape, not an individual speck of material.

THE NATURE OF OUR 'MODELS'

In this we must, of course, take shape (or *Gestalt*) in a much wider sense than as geometrical shape. *Indeed there is no observation concerned with the geometrical shape* of a particle or even of an atom. It is true that in *thinking* about the atom, in drafting theories to meet the observed facts, we do very often draw geometrical pictures on the black-board, or on a piece of paper, or more often just only in our mind, the details of the picture being given by a mathematical formula with much greater precision and in a much handier fashion than pencil or pen could ever give. That is true. But the geometrical shapes displayed in these pictures are not anything that could be directly observed in the real atoms. The pictures are only a mental help, a tool of thought, an intermediary means, from which to deduce, out of the results of experiments that have been made, a reasonable expectation about the results of new experiments that we are planning. We plan them for the purpose of seeing whether they confirm the expectations—thus whether the expectations were reasonable, and thus whether the pictures or models we use are *adequate*. Notice that we prefer to say *adequate*, not *true*. For in order that a description be *capable* of being true, it must be capable of being compared *directly* with actual facts. That is usually not the case with our models.

But we do use them, as I said, to deduce observable features from them. It is these that constitute the permanent shape or form or organization of the material object, and they have usually nothing to do with 'tiny specks of material, constituting the object'.

Take for instance the atom of iron. A very interesting and highly complicated part of its organization can be displayed again and again, whenever you like and with unalterable permanence, in the following manner. You bring a small amount of iron (or of an iron salt) into the electric arc and take a photograph of its spectrum, produced by a powerful optical grating. You find tens of thousands of sharp spectral lines, that is to say tens of thousands of definite wave-lengths contained in the light that an iron atom emits at these high temperatures. And they are always the same, exactly the same, so much so that as is well known, you can tell from the spectrum of a star that it contains certain chemical elements. While you are unable to find out anything about the geometrical shape of an atom—even with the most powerful microscope—you are able to discover its typical permanent organization, displayed in its spectrum, at a distance of thousands of light-years!

You may say the typical line spectrum of an element like iron is a macroscopic property, a property of the glowing vapour, it has nothing to do with its coarse-grained structure (its being composed of single atoms)—and nobody has yet observed the light emitted by a single, a truly isolated, atom. That is true. But, of course, I must remind you that the theory of matter, as it is accepted at present, does ascribe the emission of all these various monochromatic beams of light to the single atom; the geometrical-mechanical-electrical constitution of the single atom is deemed responsible for every single wave-length we observe in the glowing vapour. To confirm this, the physicist most emphatically points to the fact that these line spectra are only observed in the rarefied vaporous state where the atoms are so far apart from each other that they do not disturb each other. Glowing solid or liquid iron emits a continuous spectrum, much the same as every other solid or liquid at the same temperature—the sharp lines have entirely disappeared, or, better, they are entirely blurred, owing to the mutual disturbance of neighbouring atoms.

Would you then say—so you might ask me—would you then say that we are to regard the observed line spectra (which, broadly speaking, conform to the theory) as part of the *circumstantial evidence*, that the iron atoms of our theoretical description actually *exist* and that they constitute the vapour in the way the theory of gases maintains it—small specks of matter (of that peculiar constitution that makes them emit the spectral lines)—small specks of *something*, wide apart, embedded in the *nothing*, flying hither and thither, occasionally colliding with the walls, etc., etc.? Is that a *true* picture of glowing iron vapour?

I keep to what I said earlier in a more general context: it is certainly an *adequate* picture; but as regards its *truth* the appropriate question to ask is not whether it is true or not, but whether it is at all capable of being either true or false. Probably it is not. Probably we cannot ask for more than just adequate pictures capable of synthesizing in a comprehensible way all observed facts and giving a reasonable expectation on new ones we are out for.

Very similar declarations have been made again and again by competent physicists a long time ago, all through the nineteenth century and in the early days of our own. They were aware that the desire for having a *clear* picture necessarily led one to encumber it with unwarranted details. It is, so to speak, 'infinitely improbable' that those gratuitous additions should, by good luck, turn out to be 'correct'. L. Boltzmann strongly emphasized the point; let me be quite precise, he would say, childishly precise about my model, even though I know that I cannot guess from the ever incomplete circumstantial evidence of experiments what nature really is like. But without an absolutely precise model thinking itself becomes imprecise, and the consequences to be derived from the model become ambiguous.

Yet the attitude at that time—except perhaps in a very few philosophically foremost minds—was different from what it is now, it was still a little too naïve. While asserting that any model we may conceive is sure to be deficient and would surely be modified sooner or later, one still had at the back of one's mind the thought that a true model exists—exists so to speak in the Platonic realm of ideas—that we approach to it gradually, without perhaps ever reaching it, owing to human imperfections.

This attitude has now been abandoned. The failures we have experienced no longer refer to details, they are of a more general kind. We have become fully aware of a situation that may perhaps be summarized as follows. As our mental eye penetrates into smaller and smaller distances and shorter and shorter times, we find nature behaving so entirely differently from what we observe in visible and palpable bodies of our surrounding that *no* model shaped after our large-scale experiences can ever be 'true'. A completely satisfactory model *of this type* is not only practically inaccessible, but not even thinkable. Or, to be precise, we can, of course, think it, but however we think it, it is wrong; not perhaps quite as meaningless as a 'triangular circle', but much more so than a 'winged lion'.

CONTINUOUS DESCRIPTION AND CAUSALITY

I shall try to be a little clearer about this. From our experiences on a large scale, from our notion of

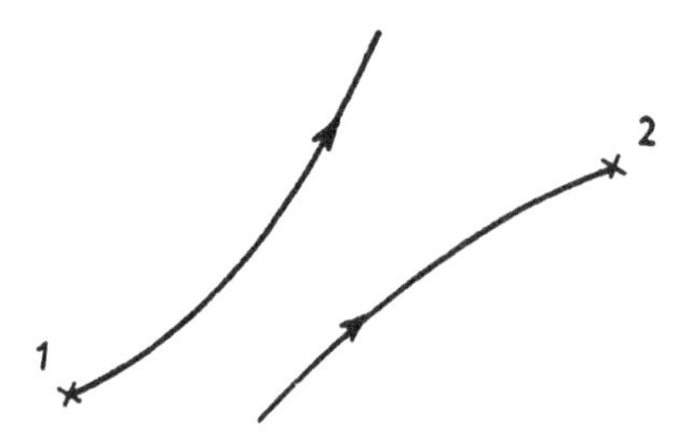

Figure 2

geometry and of mechanics—particularly the mechanics of the celestial bodies—physicists had distilled the one clear-cut demand that a truly clear and complete description of any physical happening has to fulfil: it ought to inform you precisely of what happens at any point in space at any moment of time—of course, within the spatial domain and the period of time covered by the physical events you wish to describe. We may call this demand the *postulate of continuity of the description.* It is this postulate of continuity that appears to be unfulfillable! There are, as it were, gaps in our picture.

This is intimately connected with what I called earlier the lack of individuality of a particle, or even of an atom. If I observe a particle here and now, and observe a similar one a moment later at a place very near the former place, not only cannot I be sure whether it is 'the same', but this statement has no absolute meaning. This *seems* to be absurd. For we are so used to thinking that at every moment between the two observations the first particle must have been *somewhere*, it must have followed a *path*, whether we know it or not. And similarly the second particle must have come from somewhere, it must have *been* somewhere at the moment of our first observation. So in principle it must be decided, or decidable, whether these two paths are the same or not—and thus whether it *is* the same particle. In other words we assume—following a habit of thought that applies to palpable objects—that we could have kept our particle under *continuous observation*, thereby ascertaining its identity.

This habit of thought we must dismiss. *We must not admit the possibility of continuous observation.* Observations are to be regarded as discrete, disconnected events. Between them there are gaps which we cannot fill in. There are cases where we should upset everything if we admitted the possibility of continuous observation. That is why I said it is better to regard a particle not as a permanent entity but as an instantaneous event. Sometimes these events form chains that give the illusion of permanent beings—but only in particular circumstances and only for an extremely short period of time in every single case.

Let us go back to the more general statement I made before, namely that the classical physicist's naïve ideal cannot be fulfilled, his demand that in principle information about every point in space at every moment of time should at least be *thinkable.* That this ideal breaks down has a very momentous consequence. For in the times when this ideal

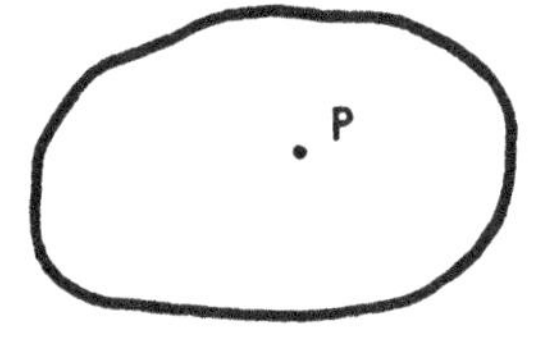

Figure 3

of continuity of description was not doubted, the physicists had used it to formulate the *principle of causality* for the purposes of their science in a very clear and precise fashion—the only one in which they could use it, the ordinary enouncements being much too ambiguous and imprecise. It includes in this form, the principle of 'close action' (or the absence of *actio in distans*) and runs as follows: The exact physical situation at *any* point P at a given moment t is unambiguously determined by the exact physical situation within a certain surrounding of P at any previous time, say t—τ. If τ is large, that is, if that previous time lies far back, it may be necessary to know the previous situation for a wide domain around P. But the 'domain of influence' becomes smaller and smaller as τ becomes smaller, and becomes infinitesimal as τ becomes infinitesimal. Or, in plain, though less precise, words: what happens anywhere at a given moment depends only and unambiguously on what has been going on in the immediate neighbourhood 'just a moment earlier'. Classical physics rested entirely on this principle. The mathematical instrument to implement it was in all cases a system of partial differential equations —so-called field equations.

Obviously, if the ideal of continuous, 'gap-less', description breaks down, this precise formulation of the principle of causality breaks down. And we must not be astonished to meet in this order of ideas with new, unprecedented difficulties as regards causation. We even meet (as you know) with the statement that there are gaps or flaws in strict causation. Whether this is the last word or not it is difficult to say. Some people believe that the question is by no manner of means settled (among them, by the way, is Albert Einstein). I shall tell you a little later about the 'emergency exit', used at present to escape from the delicate situation. For the moment I wish to attach some further remarks to the classical ideal of continuous description.

THE INTRICACY OF THE CONTINUUM

However painful its loss may be, by losing it we probably lose something that is very well worth losing. It seems simple to us, because the idea of the continuum seems simple to us. We have somehow lost sight of the difficulties it implies. That is due to a suitable conditioning in early childhood. Such an idea as 'all the numbers between 0 and 1' or 'all the numbers between 1 and 2' has become quite familiar to us. We just think of them geometrically as the distance of any point like P or Q from 0 (see fig. 4).

Among the points like Q there is also the $\sqrt{2}$ ($= 1.414 \ldots$). We are told that such a number as $\sqrt{2}$ worried Pythagoras and his school almost to exhaustion. Being used to such queer numbers from

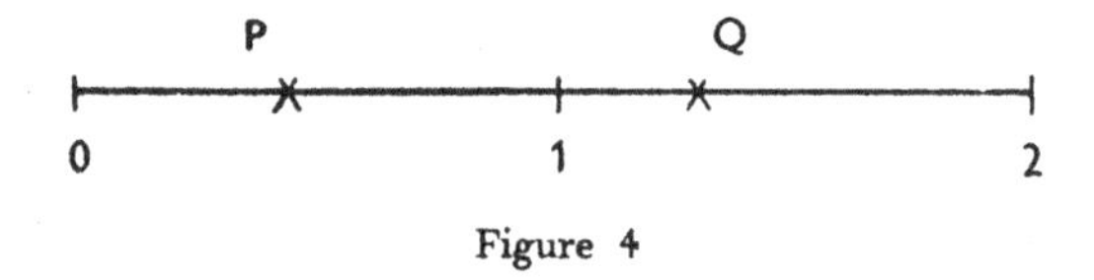

Figure 4

early childhood, we must be careful not to form a low idea of the mathematical intuition of these ancient sages. Their worry was highly creditable. They were aware of the fact that no fraction can be indicated of which the square is exactly 2. You can indicate close approximations, as for instance $\frac{17}{12}$, whose square, $\frac{289}{144}$, is very near to $\frac{288}{144}$, which is 2. You can get closer by contemplating fractions with larger numbers than 17 and 12, but you will never get *exactly* 2.

The idea of a *continuous range*, so familiar to mathematicians in our days, is something quite exorbitant, an enormous extrapolation of what is really accessible to us. The idea that you should *really* indicate the exact values of any physical quantity—temperature, density, potential, field strength, or whatever it might be—for *all* the points of a continuous range, say between zero and 1, is a bold

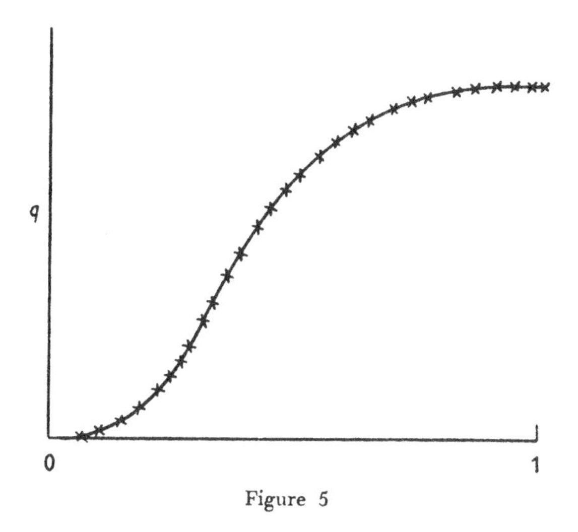

Figure 5

extrapolation. We *never* do anything else than determine the quantity approximately for a very limited number of points and then 'draw a smooth curve through them'. This serves us well for many practical purposes, but from the epistemological point of view, from the point of view of the theory of knowledge, it is totally different from a supposed exact continual description. I might add that even in classical physics there were quantities—as, for instance, temperature or density—which avowedly did not admit of an exact continuous description. But this was due to the conception these terms represent—they have, even in classical physics, only a statistical meaning. However I shall not go into details about this at the moment, it would create confusion.

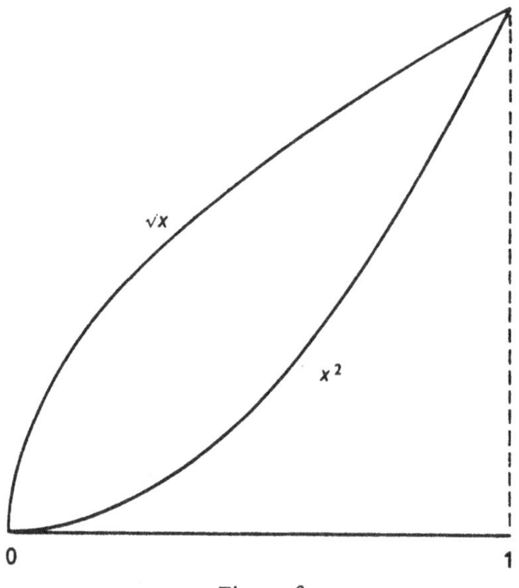

Figure 6

The demand for continuous description was encouraged by the fact that the mathematician claims to be able to indicate simple continuous descriptions of some of his simple mental constructions. For example, take again the range $0 \rightarrow 1$, call the variable in this range x, we claim to have an unambiguous idea of, say x^2 or $\sqrt{x}$.

The curves are pieces of parabolas (mirror images of each other). We claim to have full knowledge of every point of such a curve, or rather, *given* the horizontal distance (abscissa) we are able to indicate the height (ordinate) *with any required precision*. But behold the words 'given' and 'with any required precision'. The first means 'we can give the answer, when it comes to it'—we cannot possibly have all the answers in store for you in advance. The second means 'even so, we cannot as a rule give you an absolutely precise answer.' You must tell us the precision you require, e.g. up to 1000 decimal places.

$0 \quad \frac{1}{9} \quad \frac{2}{9} \quad \frac{1}{3} \qquad \frac{2}{3} \quad \frac{7}{9} \quad \frac{8}{9} \quad 1$

Figure 7

Then we can give you the answer—if you leave us time.

Physical dependences can always be approximated by this simple kind of functions (the mathematician calls them 'analytical', which means something like 'they can be analysed'). But to assume that physical dependence *is* of this simple type, is a bold epistemological step, and probably an inadmissible step.

However, the chief conceptual difficulty is the enormous number of 'answers' that are required, due to the enormous number of points contained in even the smallest continuous range. This quantity—the number of points between 0 and 1, for example—is so fabulously great that it is hardly diminished even if you take 'nearly all of them' away. Allow me to illustrate this by an impressive example.

Envisage again the line 0→1. I wish to describe a certain set of points that is *left over*, when you take some of them away, bar them, exclude them, make them inaccessible—or whatever you wish to call it. I shall use the word 'take away'.

First take away the whole middle third including its left border point, thus the points from $\frac{1}{3}$ to $\frac{2}{3}$ (but you *leave* $\frac{2}{3}$). Of the remaining two thirds you again take away the 'middle thirds', including *their* left border points, but leaving their right border points. With the remaining 'four ninths' you proceed in the same way. *And so on.*

If you actually try to continue for only a few steps you will soon get the impression that 'nothing is left over'. Indeed at every step we take away a third of the remaining length. Now supposing the Income Tax Inspector charged you first 6*s*. 8*d*. in the £, and of the remainder again 6*s*. 8*d*. in the £, and so on, *ad infinitum*, you agree you would not retain much.

We shall now analyse our case, and you will be astonished how many of our numbers or points are left. I regret that this needs a little preparation. A number between zero and one can be represented by a decadic fraction, as

$$0{\cdot}470802\ldots$$

and you know this means

$$\frac{4}{10}+\frac{7}{10^2}+\frac{0}{10^3}+\frac{8}{10^4}+\ldots.$$

That we habitually use here the number 10 is a pure

accident, due to the fact that we have 10 fingers. We can use any other number, 8, 12, 3, 2 . . . . We need, of course, different figure-symbols for all the numbers up to the chosen 'basis'. In our decadic system we need ten, 0, 1, 2, . . . 9. If we used 12 as our basis, we should have to invent single symbols for 10 and 11. If we used the basis 8, the symbols for 8 and 9 would become supernumerary.

Non-decadic fractions have not altogether been ousted by the decimal system. Dyadic fractions, that is those which use the basis 2, are quite popular, particularly with the British. When I asked my tailor the other day how much material I should get him for the flannel trousers I had just ordered, he answered—to my amazement—$1\frac{3}{8}$ yards. This is easily seen to be the *dyadic* fraction

$$1{\cdot}011,$$

meaning

$$1+\frac{0}{2}+\frac{1}{4}+\frac{1}{8}.$$

In the same way some stock exchanges quote shares not in shillings and pence but in dyadic fractions of a pound, for example £$\frac{13}{16}$, which in *dyadic* notation would read

$$0{\cdot}1101,$$

meaning

$$\frac{1}{2}+\frac{1}{4}+\frac{0}{8}+\frac{1}{16}.$$

Notice that in a dyadic fraction only two symbols, viz. 0 and 1, occur.

For our present purpose we first need *triadic*

fractions, which have the basis 3 and use only the symbols 0, 1, 2. Here, for instance, the notation

$$0{\cdot}2012\ldots$$

means

$$\frac{2}{3}+\frac{0}{9}+\frac{1}{27}+\frac{2}{81}+\ldots.$$

(By adding dots we intentionally admit fractions that run to infinity, as for example the square root of 2). Now let us return to the problem of describing the 'almost vanishing' set of numbers that is left over in the construction illustrated by our figure. A little careful thinking will shew you that the points we have *taken away* are all those which in *triadic* representation contain a figure 1 *somewhere*. Indeed, by first cutting out the middle third we cut out all the numbers whose triadic fraction begins thus:

$$0{\cdot}1\ldots.$$

At the second step we cut out all those whose triadic fraction begins

$$\text{either}\quad 0{\cdot}01\ldots \quad\text{or}\quad 0{\cdot}21\ldots.$$

And so on.—This consideration shews that there is something left, namely all those whose triadic fractions contain *no* 1, but only 0 and 2, as for instance

$$0{\cdot}22000202\ldots$$

(where the dots stand for any sequence of 0s and 2s only). Among them are, of course, the *right* border points (as $0{\cdot}2 = \frac{2}{3}$ or $0{\cdot}22 = \frac{2}{3} + \frac{2}{9} = \frac{8}{9}$) of the excluded intervals; we had decided to let those border points stand. But there are a lot more, for instance the *periodic* dyadic fraction $0{\cdot}\dot{2}\dot{0}$, meaning

0·20202020 . . . . *ad infinitum*. This is the infinite series

$$\frac{2}{3}+\frac{2}{3^3}+\frac{2}{3^5}+\frac{2}{3^7}+\ldots.$$

To find its value, think you multiply it by the square of 3, which is 9. Then the first term gives $\frac{18}{3}$, that is, 6, while the remaining terms give the same series again. Hence *eight* times our series is 6, and our number is $\frac{6}{8}$ or $\frac{3}{4}$.

Still, recalling again that the intervals we have 'taken away' tend to cover the *whole* interval between 0 and 1, one is inclined to think that, compared with the original set (containing *all* numbers between 0 and 1), the remaining set must be 'exceedingly scarce'. But now comes the amazing turn: in a certain sense the remaining set is still just as vast as the original one. Indeed we can associate their respective members in pairs, by monogamously mating, as it were, each number of the original set with a definite number of the remaining set, without any number being left over on either side (the mathematician calls this a 'one-to-one correspondence'). This is so perplexing that, I am sure, many a reader will at first think he *must* have misunderstood the words, though I have taken pains to set them as unambiguously as possible.

How is this done? Well, the 'remaining set' is represented by *all* the *triadic* fractions containing only 0s and 2s; we gave the general example

0·22000202...

(the dots standing for any sequence of 0s and 2s

only). Associate with this *triadic* fraction the *dyadic* fraction

$$0{\cdot}11000101\ldots$$

obtained from the former by replacing every figure 2 by the figure 1. Vice versa you can, from *any* dyadic fraction, by changing its 1s into 2s, obtain the *triadic* representation of a definite number in what we called 'the remaining set'. Since now any member of the original set, that is, any number between 0 and 1, is represented by one and only one[1] definite dyadic fraction, there is actually a perfect one-to-one mating between the members of the two sets.

[It may be useful to illustrate the 'mating' by examples. For instance the dyadic number that my tailor used

$$\frac{3}{8} = \frac{0}{2} + \frac{1}{4} + \frac{1}{8} = 0{\cdot}011$$

would lead to the triadic counterpart

$$0{\cdot}022 = \frac{0}{3} + \frac{2}{9} + \frac{2}{27} = \frac{8}{27};$$

that is to say, $\frac{3}{8}$ of the original set corresponds to $\frac{8}{27}$ in the remaining set. Inversely, take our triadic $0.\dot{2}\dot{0}$, meaning, as we made out, $\frac{3}{4}$. The corresponding dyadic $0.\dot{1}\dot{0}$ means the infinite series

$$\frac{1}{2} + \frac{1}{2^3} + \frac{1}{2^5} + \frac{1}{2^7} + \frac{1}{2^9} + \ldots.$$

If you multiply this by the square of 2, which is 4, you get: 2+ *the same series*. In other words, *three*

[1] We have tacitly disregarded such trivial duplications as are instanced, in the decimal system, by $0.1 = 0.0\dot{9}$ or $0.8 = 0.7\dot{9}$.

times our series equals 2, the series equals $\frac{2}{3}$; that is to say, the number $\frac{3}{4}$ of the 'remaining set' corresponds (or 'is mated') to the number $\frac{2}{3}$ in the original set.]

The remarkable fact about our 'remaining set' is that, though it covers no measurable interval, yet it has still the vast extension of any continuous range. This astonishing combination of properties is, in mathematical language, expressed by saying that our set has still the 'potency' of the continuum, although it is 'of measure zero'.

I have brought this case before you, in order to make you feel that there is something mysterious about the continuum and that we must not be all too astonished at the apparent failure of our attempts to use it for a precise description of nature.

THE MAKESHIFT OF WAVE MECHANICS

Now I shall try to give you an idea of the way in which physicists at present endeavour to overcome this failure. One might term it an 'emergency exit', though it was not intended as such, but as a new theory. I mean, of course, wave mechanics. (Eddington called it 'not a physical theory but a dodge—and a very good dodge too'.)

The situation is about as follows. The observed facts (about particles and light and all sorts of radiation and their mutual interaction) appear to be *repugnant* to the classical ideal of a continuous description in space and time. (Let me explain myself to the physicist by hinting at one example: Bohr's famous theory of spectral lines in 1913 had to assume that the atom makes a *sudden* transition from one state into another state, and that in doing so it emits a train of light waves several feet long, containing hundreds of thousands of waves and requiring for its formation a considerable time. No information about the atom during this transition can be offered.)

So the facts of observation are irreconcilable with a continuous description in space and time; it just seems impossible, at least in many cases. On the other hand, from an incomplete description—from a picture with gaps in space and time—one cannot draw clear and unambiguous conclusions; it leads to hazy, arbitrary, unclear thinking—and that is the thing we must avoid at all costs! What is to be done? The method adopted at present may seem amazing to you. It amounts to this: we do give a complete description, continuous in space and time without leaving any gaps, conforming to the classical ideal—a description *of something*. But we do not claim that this 'something' is the observed or observable facts; and still less do we claim that we thus describe what nature (matter, radiation, etc.) really *is*. In fact we use this picture (the so-called wave picture) in full knowledge that it is *neither*.

There is no gap in this picture of wave mechanics, also no gap as regards *causation*. The wave picture conforms with the classical demand for complete determinism, the mathematical method used is that of field-equations, though sometimes they are a highly generalized type of field-equations.

But what is the use of such a description, which, as I said, is not believed to describe observable facts or what nature really is like? Well, it is believed to give us *information* about observed facts and their mutual dependence. There is an optimistic view, viz. that it gives us *all* the information obtainable about observable facts and their interdependence. But this view—which may or may not be correct—is *optimistic* only inasmuch as it may flatter our pride to possess in principle all obtainable information. It is pessimistic in another respect, we might say epistemologically pessimistic. *For the information we get as regards the causal dependence of observable facts is incomplete.* (The cloven hoof must show up *somewhere*!) The gaps, eliminated from the wave picture, have withdrawn to the connection between the wave picture and the observable facts. The latter are *not* in one-to-one correspondence with the former. Plenty of ambiguity remains, and, as I said, some optimistic pessimists or pessimistic optimists believe that this ambiguity is essential, it cannot be helped.

This is the logical situation at present. I believe I have depicted it correctly, though I am quite aware that without examples the whole discussion has remained a little bloodless—just purely logical. I am also afraid that I have given you too unfavourable an impression of the wave theory of matter. I ought to amend both points. The wave theory is not of yesterday and not of 25 years ago. It made its first appearance as the wave theory of light (Huygens 1690). For the better part of 100 years[1] light waves were regarded as an incontrovertible reality, as something of which the real existence had been proved beyond all doubt by experiments on the diffraction and interference of light. I do not think that even today many physicists—certainly not experimentalists—are ready to endorse the statement

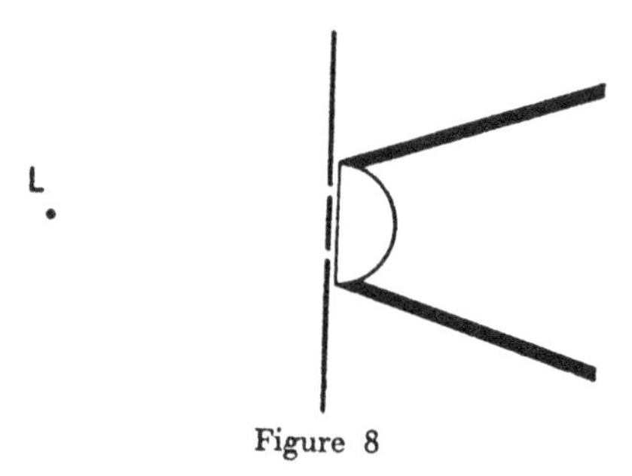

Figure 8

that 'light waves do not really exist, they are only waves of knowledge' (free quotation from Jeans).

If you observe a narrow luminous source L, a glowing Wollaston wire, a few thousandths of a millimetre thick, by a microscope whose objective lens is covered by a screen with a couple of parallel slits, you find (in the image plane conjugate to L) a system of coloured fringes which conform exactly and quantitively to the idea that light of a given colour is a wave motion of a certain small wave-

[1] Not the immediately following hundred years. Newton's authority eclipsed Huygens' theory for about a century.

length, shortest for violet, about twice as long for red light. This is one out of dozens of experiments that clinch the same view. Why, then, has this *reality* of the waves become doubtful? For two reasons:

(a) Similar experiments have been performed with beams of cathode rays (instead of light); and cathode rays—so it is said—*manifestly* consist of single electrons, which yield 'tracks' in the Wilson cloud chamber.

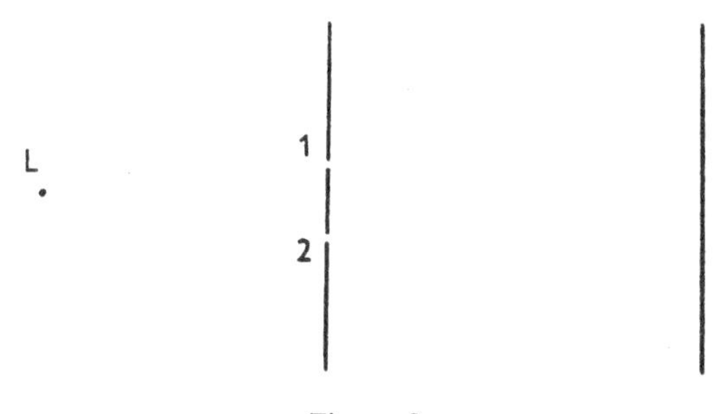

Figure 9

(b) There are reasons to assume that light itself also consists of single particles—called photons (from the Greek φῶς = light).

Against this one may argue that nevertheless in *both* cases the concept of waves is unavoidable, if you wish to account for the interference fringes. And one may also argue that the particles are not identifiable objects, they might be regarded as explosion like events within the wave-front—just the events by which the wave-front manifests itself to observation. These events—so one might say—are to a certain extent fortuitous, and that is why there is no strict causal connection between observations.

Let me explain in some detail why the phenomena, both in the case of light and in the case of cathode rays, cannot possibly be understood by the concept of single, individual, *permanently existing* corpuscles. This will also afford an example of what I call the 'gaps' in our description and of what I call the 'lack of individuality' of the particles. For the sake of

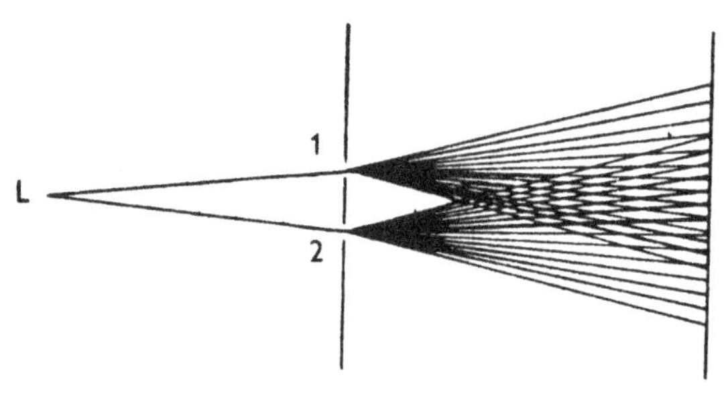

Figure 10

argument we simplify the experimental arrangement to the utmost. We consider a small, almost point-like source which emits corpuscles in all directions, and a screen with two small holes, with shutters, so that we can open first only the one, then only the other, then both. Behind the screen we have a photographic plate which collects the corpuscles that emerge from the openings. After the plate has been developed, it shows, let me assume, the marks of the single corpuscles that have hit it, each rendering a grain of silver-bromide developable, so that it shows as a black speck after developing. (This is very near the truth.)

Now let us first open only one hole. You might expect that after exposing for some time we get a close cluster around one spot. This is not so. Apparently the particles are deflected from their straight

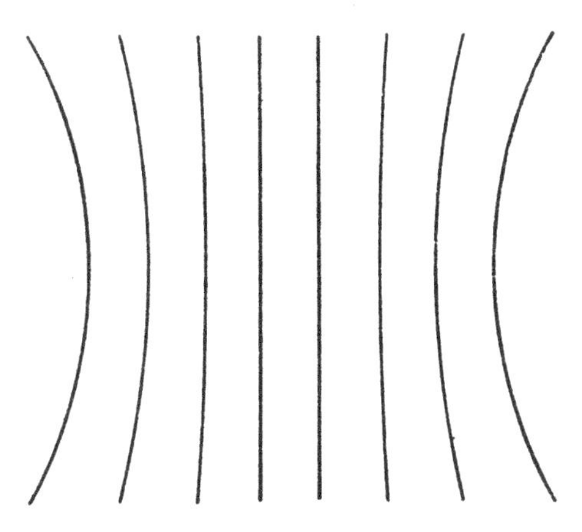

Figure 11

The lines indicate the places where there are few or no spots, while midway between any two lines the spots would be most frequent. The two straight lines in the middle are parallel to the slits.

path at the opening. You get a fairly wide spreading of black specks, though they are densest in the middle, becoming rarer at greater angles. If you open the second hole alone, you clearly get a similar pattern, only around a different centre.

Now let us open both holes at the same time and expose the plate just as long as before. What would

you expect—if the idea was correct, that single individual particles fly from the source to one of the holes, are deflected there, then continue along another straight line until they are caught by the plate? Clearly you would expect to get the two former patterns superposed. Thus in the region where the two fans overlap, if near a given point of the pattern you had, say, 25 spots per unit area in the first experiment and 16 more in the second, you would expect to find $25 + 16 = 41$ in the third experiment. This is not so. Keeping to these numbers (and *disregarding chance-fluctuations*, for the sake of argument), you may find anything between 81 and only 1 spot, this depending on the precise place on the plate. It is decided by the difference of its distances from the holes. The result is that in the overlapping part we get dark fringes separated by fringes of scarcity.
(N.B. The numbers 1 and 81 are obtained as

$$(\sqrt{25} \pm \sqrt{16})^2 = (5 \pm 4)^2 = \frac{81}{1}. \;)$$

If one wanted to keep up the idea of single individual particles flying continuously and independently either through one or through the other slit one would have to assume something quite ridiculous, namely that in some places on the plate the particles destroy each other to a large extent, while at other places they 'produce offspring'. This is not only ridiculous but can be refuted by experiment. (Making the source extremely weak and exposing for a very long time. This does not change the pattern!) The only other alternative is to assume that a particle flying through the opening No. 1 is influenced also by the opening No. 2, and that in an extremely mysterious fashion.

We must, so it seems, give up the idea of tracing back to the source the history of a particle that manifests itself on the plate by reducing a grain of silver-bromide. *We cannot tell where the particle was before it hit the plate.* We cannot tell through which opening it has come. This is one of the typical gaps in the description of observable events, and very characteristic of the lack of individuality in the particle. We must *think* in terms of spherical waves emitted by the source, parts of each wave-front passing through both openings, and producing our interference pattern on the plate—but this pattern manifests itself to *observation* in the form of single particles.

THE ALLEGED BREAK-DOWN OF THE BARRIER BETWEEN SUBJECT AND OBJECT

It cannot be denied that the new physical aspect of nature of which I have tried to give you some idea by this example is very much more complicated than the old way which I called 'the classical ideal of uninterrupted, continuous description'. The very serious question arises naturally: Is this new and unfamiliar way of looking at things, which is at variance with the habits of everyday thinking—is it deeply rooted in the facts of observation, so that it has *come to stay* and will never by got rid of again; or is this new aspect perhaps the mark, not of objective nature, but of the setting of the human mind, of the stage that our understanding of nature has reached at present?

This is an extremely difficult question to answer, because it is not even absolutely clear what this antithesis means: objective nature and human mind. For on the one hand I undoubtedly form part of nature, while on the other hand objective nature is known to me as a phenomenon of my mind only. Another point that we must keep in mind in pondering this question is this: that one is very easily deceived into regarding an acquired habit of thought as a peremptory postulate imposed by our mind on any theory of the physical world. The famous instance of this is Kant, who, as you know, termed *space* and *time, as he knew them*, the form of our mental intuition (*Anschauung*)—space being the form of external, time that of internal, intuition. Throughout the nineteenth century most philosophers followed him in this. I will not say that Kant's idea was completely wrong, but it was certainly too rigid and needed modification when new possibilities came to light, e.g. that space may be (and probably is) closed in itself, yet without boundaries; and that two events may happen in such a way that *either of them* may be regarded as the earlier one (this was the most amazing novel aspect in Einstein's 'Restricted' Theory of Relativity).

But let us return to our question, however poorly

it may be formulated: Is the impossibility of a continuous, gapless, uninterrupted description in space and time really founded in incontrovertible facts? The current opinion among physicists is, that this *is* the case. Bohr and Heisenberg have put forward a very ingenious theory about it, which is so easy to explain that it has entered most popular treatises on the subject—I should say, unfortunately; for its philosophical implication is usually misunderstood. I am going to argue against it, but first I must summarize it briefly.

It runs as follows. We cannot make any factual statement about a given natural object (or physical system) without 'getting in touch' with it. This 'touch' is a real physical interaction. Even if it consists only in 'looking at the object', the latter must be hit by light-rays and reflect them into the eye, or into some instrument of observation. This means that the object is *interfered with* by observing it. You cannot obtain any knowledge about an object while leaving it strictly isolated. The theory goes on to assert that this disturbance is neither irrelevant nor completely surveyable. Thus after any number of painstaking observations the object is left in a state of which *some* features (the last observed ones) are known, but *others* (those interfered with by the last observation) are not known, or not accurately known. This state of affairs is offered as the explanation why no complete, gapless description of a physical object is possible.

But obviously these inferences, even when granted,

tell me so far only that such a description cannot be actually accomplished, but they do *not* convince me that I should not be able to form *in my mind* a complete, gapless *model*, from which everything I can observe can be correctly inferred or foreseen, to the degree of certainty which the incompleteness of my observations allows. The situation *might* be such as in the beginning of a game of whist. By the rules of the game I can only have knowledge of one quarter of all the 52 cards. Still I know that each of the other players also has a certain lot of 13 cards, which will not change during the game; that nobody else can have a queen of hearts (because I have it); that there are exactly 6 clubs among the cards I do not know (because I happen to have 7)—and so on.

I say this interpretation suggests itself: that there *is* a fully determined physical object in existence, but I can never know all about it. However, this would be a complete misunderstanding of what Bohr and Heisenberg and those who follow them actually mean. They mean that the object has no existence independent of the observing subject. They mean that recent discoveries in physics have pushed forward to the mysterious boundary between the *subject* and the *object*, which thereby has turned out not to be a sharp boundary at all. We are to understand that we never observe an object without its being modified or tinged by our own activity in observing it. We are to understand that under the impact of our refined methods of observation, and of thinking about the results of our experiments, that mysterious boundary between the subject and the object *has broken down.*

The opinion of what may be called our two foremost quantum theorists deserves, of course, careful attention; and the further fact that several other prominent scientists do not reject their opinion, but seem rather satisfied with it, adds to its claim to be thoroughly weighed. But in doing so, I cannot suppress certain objections.

I do not think I am prejudiced against the importance that science has from the purely human point of view. I had expressed by the original title of these lectures, and I have explained in the introductory passages, that I consider science an integrating part of our endeavour to answer the one great philosophical question which embraces all others, the one that Plotinus expressed by his brief: τίνες δὲ ἡμεῖς; —*who are we*? And more than that: I consider this not only one of the tasks, but *the* task, of science, the only one that really counts.

But with all that, I cannot believe (and this is my first objection)—I cannot believe that the deep philosophical enquiry into the relation between subject and object and into the true meaning of the distinction between them depends on the quantitative results of physical and chemical measurements with weighing scales, spectroscopes, microscopes, telescopes, with Geiger-Müller-counters, Wilson-chambers, photographic plates, arrangements for measuring the radioactive decay, and whatnot. It is not very easy to say *why* I do not believe it. I feel a certain incongruity between the applied means and the problem to be solved. I do *not* feel quite so diffident with regard to other sciences, in particular biology, and quite especially *genetics* and the facts about *evolution*. But we shall not talk about this here and now.

On the other hand (and this is my second objection), the mere contention that every observation depends on both the subject and the object, which are inextricably interwoven—this contention is hardly new, it is almost as old as science itself. Though but scarce reports and quotations of the two great men from Abdera, Protagoras and Democritus, have come down to us across the twenty-four centuries that separate us from them, we can tell that they both in their way maintained that all our sensations, perceptions, and observations have a strong personal, subjective tinge and do not convey the nature of the thing-in-itself (the difference between them was that Protagoras dispensed with the thing-in-itself, to him our sensations were the only truth, while Democritus thought differently). Since then the question has turned up whenever there was science; we might follow it through the centuries, speaking of Descartes', Leibnitz', Kant's attitudes towards it. We shall not do this. But I must mention one point, in order not to be accused of injustice towards the quantum physicists of our days. I said their statement that in perception and observation subject and object are inextricably interwoven is hardly new.

But they could make a case that something about it *is* new. I think it is true that in previous centuries, when discussing this question, one mostly had in mind two things, viz. (a) a direct physical *impression caused* by the object in the subject, and (b) the *state* of the subject that receives the impression. As against this, in the present order of ideas the direct physical, causal, influence between the two is regarded as *mutual*. It is said that there is also an unavoidable and uncontrollable impression from the side of the *subject* onto the *object*. This aspect *is* new, and, I should say, more adequate anyhow. For physical action always is *inter*-action, it always *is* mutual. What remains doubtful to me is only just this: whether it is adequate to term one of the two physically interacting systems the 'subject'. *For the observing mind is not a physical system, it cannot interact with any physical system.* And it might be better to reserve the term 'subject' for the observing mind.

ATOMS OR QUANTA—THE COUNTER-SPELL OF OLD STANDING, TO ESCAPE THE INTRICACY OF THE CONTINUUM

Be this as it may, it seems worth our while to try to examine the matter from various angles. A point of view that I have previously touched on in these lectures and that does suggest itself is this, that our present difficulties in physical science are bound up with the notorious conceptional intricacy inherent in the idea of the *continuum*. But this does not tell you much. How are they bound up? What precisely is the mutual relationship?

If you envisage the development of physics in *the last half-century*, you get the impression that the discontinuous aspect of nature has been forced upon us *very much against our will*. We seemed to feel quite happy with the continuum. Max Planck was seriously frightened by the idea of a discontinuous exchange of energy, which he had introduced (1900) in order to explain the distribution of energy in black-body-radiation. He made strong efforts to weaken the hypothesis, and, if possible, to get away from it, but in vain. Twenty-five years later the inventors of wave mechanics indulged for some time in the fond hope that they had paved the way of return to a classical continuous description, but again the hope was deceptive. Nature herself seemed to reject continuous description, and this refusal seemed to have *nothing* to do with the mathematicians' *aporia* in dealing with the continuum.

This is the impression you get from the last 50 years. But quantum theory dates 24 centuries further back, to Leucippus and Democritus. They invented the first discontinuity—isolated atoms embedded in empty space. Our notion of the elementary particle has historically descended from their notion of the atom and is conceptionally derived from their notion of the atom; *we have simply held on to it*. And these *particles* have now turned out to be *quanta of energy*, because—as Einstein discovered in 1905—

mass and energy are the same thing. So the idea of discontinuity is very old. How did it arise? I wish to establish that it originated precisely from the intricacy of the continuum, so to speak as a weapon in defence against it.

How did the ancient atomists come by the idea of atomism of matter? This question gains now a more than merely historical interest, it becomes epistemologically relevant. The question is sometimes asked in the following form—in a mood of utter amazement: How did those thinkers, with an extremely scanty knowledge of the laws of physics, indeed in complete ignorance of all the relevant experimental facts—how did they hit on the *correct* theory of the composition of material bodies? Occasionally you find people so bewildered by this 'lucky strike' that they actually declare it to be a chance-event and refuse to give the ancient atomists any credit for it. They declare that their atomic theory has been a completely unfounded guess which might just as well have turned out a mistake. Needless to say, it is always a scientist, never a classical scholar, who reaches this strange conclusion.

I reject it. But then I must answer the question. That is not very difficult. The atomists and their ideas did not emerge suddenly out of nothing, they were preceded by the great development that began with Thales of Miletus (floruit 585 B.C.) more than a century earlier; they continue the awe-inspiring line of Ionian physiologoi. Their immediate predecessor in this line was Anaximenes, whose principal doctrine consisted in underlining the all-importance of 'rarefaction and condensation'. From a careful consideration of everyday experience he abstracted the thesis that every piece of matter can take on the solid, the liquid, the gaseous and the 'fiery' state; that the changes between these states do not imply a change of nature, but are brought about geometrically, as it were, by the spreading of the same amount of matter over a larger and larger volume (rarefaction), or—in the opposite transitions—by its being reduced or compressed into a smaller and smaller volume. This idea is so absolutely to the point that a modern introduction into physical science could take it over without any relevant change. Moreover it is certainly not an unfounded guess, but the outcome of careful observation.

If you try to assimilate Anaximenes' idea, you naturally come to think that the change of properties of matter, say on rarefaction, must be caused by its parts receding at greater distances from each other. But it is extremely difficult to accomplish this in your imagination, if you think of matter as forming a gapless continuum. What should recede from what? The mathematicians of the same epoch considered a geometrical line as consisting of points. That is perhaps all right if you leave it alone. But if it is a *material* line and you begin to stretch it—would not its points recede from each other and leave gaps between them? For the stretching cannot *produce* new points and the same set of points cannot go to cover a greater interval.

From these difficulties, *which reside in the mysterious character of the continuum*, the easiest escape is the one taken by the atomists, namely to regard matter as consisting from the outset of isolated 'points' or rather of small particles, which recede from each other on rarefaction and approach to closer distances on condensation, while remaining themselves unchanged. The latter is an important by-product. Without it, the contention that in these processes matter stays intrinsically unchanged would remain very hazy. The atomist can tell what it means: the particles remain unchanged; only their geometrical constellation changes.

It would thus seem that physical science in its present form—in which it is the direct offspring, the uninterrupted continuation, of ancient science—was from its very beginning ushered in by the desire to avoid the haziness inherent in the conception of the continuum, the precarious side of which was then more felt than in modern times, until quite recently. Our helplessness vis-à-vis the continuum, reflected in the present difficulties of quantum theory, is not a late arrival, it stood godmother to the birth of science—an evil godmother, if you please, like the thirteenth fairy in the tale of the Sleeping Beauty. Her evil spell had for a long time been stemmed by the genial invention of atomism. *This explains why atomism has proved so successful and durable and indispensable.* It was not a happy guess by thinkers who 'really did not know anything about it'—it was the powerful counter-spell which naturally cannot be dispensed with as long as the difficulty it is to exorcise survives.

By this I will not say that atomism will ever go by the board. Its invaluable findings—especially the statistical theory of heat—certainly never will. Nobody can tell the future. Atomism finds itself facing a serious crisis. Atoms—our modern atoms, the ultimate particles—must no longer be regarded as identifiable individuals. This is a stronger deviation from the original idea of an atom than anybody had ever contemplated. We must be prepared for anything.

WOULD PHYSICAL INDETERMINACY GIVE FREE WILL A CHANCE?

On p. 12 I briefly touched upon that old crux, the apparent contradiction between the deterministic view about material events and what is called in Latin *liberum arbitrium indifferentiae*, in modern language free will. I suppose you all know what I mean: since my mental life is obviously bound up very closely with the physiological goings on in my body, more especially in my brain, then, if the latter are strictly and uniquely determined by physical and chemical natural laws, what about my inalienable feeling that *I* take decisions to act in this or that way, what about my feeling responsibility for the decision I actually do take? Is not everything I do mechanically determined in advance by the material state of affairs in my brain, including modifications caused by external bodies, and is not my feeling of liberty and responsibility deceptive?

This does strike us as a true *aporia*, which occurred for the first time to Democritus, who realised it fully —but left it alone; very wisely, I think. He fully realised it. While he adhered to his 'atoms and the void' as the only reasonable way of understanding objective nature, we have some definite utterances of his preserved, to the effect that he also realised that this whole picture of the atoms and the void was formed by the human mind on the evidence of sense perceptions, and nothing else; and other utterances where he states, almost in the words of Kant, that we know nothing about what anything really is in itself, the ultimate truth remaining deeply in the dark.

Epicurus took over Democritus' physical theories (by the way, without acknowledgement); however, less wise, and very keen on conveying to his disciples a fair and sound and incontrovertible *moral* attitude, he tampered with physics and invented his famous (or ill-famed) swerves—strongly reminiscent of modern ideas about 'uncertainty' of physical events. I will not enter on details here; suffice it to say that he broke away from physical determinism in a rather childish way, which was not based on any experience and therefore had no consequences.

The problem itself never left us. It turned up very prominently—or at least a problem of closely similar *logical* structure turned up—with St Augustine of Hippo, as a theological *aporia*. The part of the Law of Nature is taken by the omniscient and almighty God. But since to him who believes in God the Law of Nature is obviously His law, I think I am right in calling it very much the same problem.

As everybody knows, St Augustine's great difficulty was precisely this: God being omniscient and almighty, I cannot do a thing without His knowing and willing—not only consenting, but determining it. How, then, could I be responsible for it? I suppose the religious attitude to this form of the question eventually has to be that we are here confronted with a deep mystery into which we cannot penetrate, but which we certainly must not try to solve by denying responsibility. We must not try, I say; or *we had better not* try, for we fail pitiably. The feeling of responsibility is congenital, nobody can discard it.

But let us turn to the original form of the question and to the part physical determinism plays in it. Naturally the so-called 'crisis of causality' in the physics of our day seemed to raise strong hope of releasing us from this paradox or *aporia*.

Could perhaps the declared *indeterminacy* allow *free will* to step into the gap in the way that *free will determines* those events which the Law of Nature leaves undetermined? This hope is, at first sight, obvious and understandable.

In this crude form the attempt was made, and the idea, to a certain extent, worked out by the German physicist Pascual Jordan. I believe it to be both physically and morally an impossible solution. As regards the first: according to our present view the quantum laws, though they leave the single event undetermined, predict a quite definite *statistics* of events when the same situation occurs again and again. If these statistics are interfered with by any agent, this agent violates the laws of quantum mechanics just as objectionably as if it interfered —in pre-quantum physics—with a strictly causal mechanical law. Now we know that *there is no statistics* in the reaction of the same person to precisely the same moral situation—the rule is that the same individual in the same situation acts again precisely in the same manner. (Mind you, in *precisely* the same situation; this does not mean that a criminal or addict cannot be converted or healed by persuasion and example or whatnot—by strong external influence; but this, of course, means that the situation is changed.) The inference is that Jordan's assumption—the direct stepping in of free will to fill the gap of indeterminacy—does amount to an interference with the laws of nature, even in their form accepted in quantum theory. But at *that* price, of course, we can have everything. This is not a solution of the dilemma.

The moral objection was strongly emphasized by the German philosopher Ernst Cassirer (who died in 1945 in New York as an exile from Nazi Germany). Cassirer's extended criticism of Jordan's ideas is based on a thorough familiarity with the situation in physics. I shall try to summarize it briefly; I would say, it amounts to this. Free will in man includes as its most relevant part man's ethical behaviour. Supposing the physical events in space and time actually are to a large extent not strictly determined but subject to pure chance, as most physicists in our time believe, then this haphazard side of the goings-on in the material world is certainly (says Cassirer) *the very last to be invoked as the physical correlate of man's ethical behaviour.* For this is anything but haphazard, it is intensely determined by motives ranging from the lowest to the most sublime sort, from greed and spite to genuine love of the fellow creature or sincere religious devotion. Cassirer's lucid discussion makes one feel so strongly the absurdity of basing free will, including ethics, on physical haphazard that the previous difficulty, the antagonism between free will and determinism, dwindles and almost vanishes under the mighty blows Cassirer deals to the opposite view. 'Even the reduced extent of predictability' (Cassirer adds) 'still granted by Quantum Mechanics, would amply suffice to destroy ethical freedom, if the concept and true meaning of the latter were irreconcilable with predictability'. Indeed, one begins to wonder whether the supposed paradox is really so shocking, and whether physical determinism is not perhaps quite a suitable correlate to the mental phenomenon of will, which is not always easy to predict 'from outside', but usually extremely determined 'from inside'. To my mind this is the most valuable outcome of the whole controversy: the scale is turned in favour of a possible reconciliation of free will with physical determinism, when we realise how inadequate a basis physical haphazard provides for ethics. One could enlarge on this point. Innumerable passages could be adduced from poets and novelists to clinch it. In John Galsworthy's novel *The Dark Flower* (Part I, *13*, second paragraph) the scattered thoughts of a young lad at night hit on this: 'But that was it—you never could think what things would be like if they weren't just what and where they were. You never knew what was coming, either; and yet, when it came, it seemed as if nothing else ever could have come. That was queer—you could do anything you liked until you'd done it, but when you *had* done it then you knew, of course, that you must always have had to . . . ' There is a famous passage in *Wallenstein's Tod* (II.3):

> Des Menschen Taten und Gedanken, wisst!
> Sind nicht wie Meeres blindbewegte Wellen.
> Die innre Welt, sein Mikrokosmus, ist
> Der tiefe Schacht, aus dem sie ewig quellen.
> Sie sind notwendig, wie des Baumes Frucht;
> Sie kann der Zufall gaukelnd nicht verwandeln.
> Hab' ich des Menschen Kern erst untersucht,
> So weiss ich auch sein Wollen und sein Handeln.

> Be ye aware: man's thinking and man's deeds
> Are not like the ocean's blindly surging spray.
> His inner world, his microcosmus, feeds
> The profound shaft from which they pour to the day.
> They are needful as its fruit is in a tree,
> Unalterable by blindly juggling chance.
> Once into a man's deep core I probing see,
> His will and act I'll tell you in advance.

It is true that in their context these lines refer to

Wallenstein's devout belief in astrology, which we are not inclined to share. But is not the very lure of astrology, the irresistible attraction it has for scores of centuries exerted on men's minds, witness to the fact that we are not prepared to regard our fate as the outcome of pure chance, even though, or rather just because, it largely depends on our taking the right decision in the right moment? (We usually lack the full information needed for this purpose; and that is where astrology comes in!)

THE BAR TO PREDICTION, ACCORDING TO NIELS BOHR

But let us return to our subject proper. A much more serious and interesting attempt to explain the difficulty away was founded by Bohr and Heisenberg on the idea, mentioned above, that there is an unavoidable and uncontrollable mutual interaction between the observer and the observed physical object. Their ratiocination is briefly as follows. The alleged paradox consists in this, that according to the mechanistic view, by procuring an exact knowledge of the configuration and velocities of all the elementary particles in a man's body, including his brain, one could predict his voluntary actions—which thereby cease to be what he believes them to be, namely voluntary. The fact that we cannot *actually* procure this detailed knowledge is no great help. Even the theoretical predictability shocks us.

To this Bohr answers that the knowledge cannot

even be procured *in principle*, not even in theory, because such accurate observation would involve so strong an interference with 'the object' (the man's body) as to dissociate it into single particles—in fact kill him so efficiently that not even a corpse would be left for burial. At any rate, no prediction of behaviour would result, before the 'object' is far beyond the state of exhibiting any voluntary behaviour.

The emphasis is of course on the phrase 'in principle'. That the said knowledge cannot *actually* be procured, not even for the simplest living organism, let alone a higher animal like man, is clear also without quantum theory and uncertainty relation.

Bohr's consideration is no doubt interesting. Yet, I should say, we are more convicted by it than convinced, as in some mathematical proofs: you must grant A and B, then follows C and D, and so on, you cannot object to any single step; finally follows the interesting result Z. You have to accept it, but you cannot see how it really comes about, the proof gives no hint of that. In the present case I would say: Bohr's considerations show you that the present views in physics—mainly on account of the lack of strict causality (or on account of the uncertainty relation) —bar the objectionable predictability *in principle*. But you cannot see how this comes about. In view of the close relation Bohr's reasoning has to the lack of observable strict causality, you even incline to suspect that it is only Jordan's suggestion over again, but in a more careful disguise, so as to be shielded from Cassirer's arguments.

One can make a case for this being so. Indeed, I think I must accuse Bohr—though in actual fact he is one of the kindliest persons I ever came to know—of an unnecessary cruelty for his proposing to kill his victim by observation. I cannot see what purpose it should serve. It will never, according to quantum mechanics, yield us the full set of configuration *and* velocities of all the particles, because according to our present views this is impossible. The equivalent of this *complete* knowledge in classical physics is in quantum physics a so-called maximum observation, which yields the maximum knowledge that can be obtained, nay, that has any meaning. *Nothing in the views accepted at present precludes that we should obtain this maximum knowledge of a living body.* We must admit the possibility *in principle*, even though we know perfectly well that practically it cannot be achieved. This state of affairs is exactly the same as with *complete* knowledge in classical physics. Furthermore, precisely as in classical physics, you can from a maximum observation, yielding maximum knowledge *now*, deduce, *in principle*, maximum knowledge at any later time. (You must, of course, procure maximum knowledge also about all agents that act on your object in the meantime; but that is, in principle, possible and is again absolutely analogous to the case of classical mechanistic physics.) The fundamental difference is only this, that the said maximum knowledge at that later time may leave you in doubt about very conspicuous features of the actual observable behaviour of your object at that later time—the more so, the longer the time that has elapsed.

It would thus appear that Bohr's considerations adduce a *physical* unpredictability of the behaviour of a living body again precisely from the lack of strict causation, maintained by quantum theory. Whether or no this physical indeterminacy plays any relevant role in organic life, we must, I think, sternly refuse to make it the physical counterpart of voluntary actions of living beings, for the reasons outlined before.

The net result is that quantum physics has nothing to do with the free-will problem. If there is such a problem, it is not furthered a whit by the latest development in physics. To quote Ernst Cassirer again: 'Thus it is clear . . . that a possible change in the physical concept of causality can have no immediate bearing on ethics'.

LITERATURE

A. S. Eddington, *The Nature of the Physical World* (Gifford Lectures 1927). Cambridge University Press, 1929.

Ernst Cassirer, *Determinismus und Indeterminismus in der modernen Physik*. Götheborgs Högskolas Arsskrift *42*; Götheborg, 1937.

Pascual Jordan, *Anschauliche Quantentheorie*. Springer, Berlin, 1936.

N. Bohr, 'Licht und Leber', *Naturw*. *21*, 245, 1933.

W. Heisenberg, *Wandlungen in den Grundlagen der Naturwissenschaft*. S. Hirzel, Leipzig, 1935-1947.

M. Born, *Natural Philosophy of Cause and Chance*. Oxford University Press, 1949.

Volume VII of the 'Library of Living Philosophers', *Albert Einstein: Philosopher-Scientist*. (A collective volume, concluded by a critical essay of Einstein's, an excerpt of which is reprinted in 'Physics Today', February 1950.)

Hermann Diels, *Die Fragmente der Vorsokratiker*. Weidmann'sche Buchhandlung, Berlin, 1903.

E. C. Titchmarsh, *Theory of Functions*. Oxford University Press, 1939.

José Ortega y Gasset, *La rebelión de las masas*. Espasa-Calpe Argentina, Buenos Aires—Mexico, 1937. (This edition is enhanced by a 'Prologue for Frenchmen' and an 'Epilogue for Englishmen'. There are translations of the book in English, French and German.)